Non-Neutral Evolution

Theories and Molecular Data

Non-Neutral Evolution

Theories and Molecular Data

Edited by
Brian Golding
Department of Biology
McMaster, University

The Canadian Institute for Advanced Research
L'Institut canadien de recherches avancees

CHAPMAN & HALL
I(T)P An International Thomson Publishing Company

New York • Albany • Bonn • Boston • Cincinnati
• Detroit • London • Madrid • Melbourne • Mexico City
• Pacific Grove • Paris • San Francisco • Singapore
• Tokyo • Toronto • Washington

First published in 1994 by
Chapman & Hall
One Penn Plaza
New York, NY 10119

Published in Great Britain by
Chapman & Hall
2-6 Boundary Row
London SE1 8HN

Printed in the United States of America

Library of Congress Cataloging in Publication Data

Non-neutral evolution : theories and molecular data / editor, Brian Golding.
p. cm.
Papers from a workshop sponsored by the Canadian Institute for Advanced Research.
Includes bibliographical references and index.
ISBN 0-412-05391-8
1. Molecular evolution—Congresses. 2. Molecular genetics—Congresses. 3. Population genetics—Congresses. I. Golding, Brian, 1953– . II. Canadian Institute for Advanced Research.
QH371.N66 1994
574.87′3224—dc20

93-47006
CIP

British Library Cataloguing in Publication Data available

Contents

Preface

Are molecular differences between species adaptive? Are differences within species adaptive? These questions are central to understanding how evolution occurs. This collection of papers is a result of a workshop, sponsored by the Canadian Institute for Advanced Research, to examine these questions. The workshop attracted some of the most prominent scientists working in this area, and they present here their novel data about the molecular structure of populations and about the nature of the differences between species.

Evolution is one of the basic features of all life, and all organisms must evolve to survive—from the AIDS virus, to bacteria, to fish, to humans. But despite the central place of evolution within biology, there are many aspects of evolution that are still poorly understood. For Charles Darwin, the driving force behind all evolution was natural selection; the differential ability of organisms with different genetic compositions to survive and to leave offspring.

In 1968 Motoo Kimura and in 1969 Jack King and Thomas Jukes proposed quite a different scenario for molecular evolution. They proposed that most of the molecular changes that occur within and between organisms are neutral with respect to natural selection. They suggested that these changes are neither selectively advantageous nor deleterious to the organism. These changes were thought to produce an organism with genes that may be different from the rest of the population but which are still completely capable of carrying out their function. They did not suggest that none of the changes could be influenced by natural selection, but rather that these will be a minor component of the many other neutral changes that are possible. This became known as the **neutral theory of molecular evolution** and was subsequently skillfully championed by Motoo Kimura.

The biotechnology that was developed in the late 1980s now permits us to identify precisely the actual DNA structure from many individuals within a population. This has lead to a flood of DNA sequence data that will have many implications for the future of biology. Early studies showed that there is a large amount of variation between different DNA sequences sampled from a population. These new data allow more precise answers to questions that deal with how these DNA sequences change over time. At the same time, they pose a challenge to our ability to understand the observed patterns.

In addition to technological advances, there have also been theoretical advances. It has been shown that many evolutionary processes with or without natural selection may lead to the same kinds of patterns as does a simple model of neutral genetic variation. Hence, our ability to discern the difference between neutral genetic variation and naturally selected variation may not have been as great as we once thought.

The accumulated DNA sequence data seem to have many features that are not predicted by simplified, perhaps oversimplified, neutral models. In particular, the data show that molecular sequences evolve with bursts of changes rather than at a constant rate. But it is not simple to decide if these changes are due to natural selection or perhaps due to some simplification of the neutral models of genetic change. Generally, the lengths of time separating species have been many tens of millions of years, and, within this time, it is conceivable that many fundamental processes affecting the biology of these organisms could have changed.

By combining an examination of genetic variation among closely related species with a study of genetic variation within species, these experiments take on new power. These studies have the advantage of the precise identification of the genetic composition of each individual from their DNA sequence. In addition, intermixed within those parts of the DNA that determine the structure of proteins are other parts that, if altered, do not affect the structure of proteins. John McDonald and Marty Kreitman (1992) found that these two parts show different patterns of change. They found an excess of fixed differences between species that altered proteins while most differences within species were due to changes that did not alter proteins. These methods of comparing polymorphism and divergence provide a powerful way to examine the forces that shape molecular evolution.

The nature of the biological problems are outlined in the first few papers. John Gillespie examines several patterns of adaptive evolutionary change to determine if they can account for the bursts of evolutionary changes. He shows that even complicated models cannot account for the patterns unless evolutionary parameters such as mutation rates change over long time scales. Walter Eanes contrasts the differences of the within and between species variation in several *Drosophila* genes. He shows that the phenomena observed by McDonald and Kreitman are also present for the sequence encoding glucose-6-phosphate dehydrogenase (G6PD). Four other genes involved in central metabolism do not show this unusual pattern, and Eanes discusses why G6PD might show this pattern. The papers by Steve Schaeffer; Charles Aquadro, David Begun, and Eric Kindahl;

Wolfgang Stephan; and Montserrat Aguadé and Chuck Langley illustrate the amazing detail revealed by sequence data and show that the current theories available today cannot adequately explain these data. Schaeffer reviews the mass of sequence data he has collected from the *Adh* gene of *Drosophila pseudoobscura*. Aquadro et al. show that rates of recombination vary across the genome and that levels of nucleotide diversity covary with recombination. But they show that this is not the case with percent divergence and discusses this phenomenon. Stephan demonstrates adaptive selection from an examination of gene regions with high versus low levels of recombination in *D. ananassae*. Aguadé and Langley discuss potential causes for the large changes in levels of genetic polymorphism correlated with levels of recombination.

A major advance prompted by the neutral theory is that adaptive evolution must now be justified rigorously and cannot be simply assumed. But to do this is not an easy task when the patterns of adaptive evolution and neutral evolution can be so similar. The papers by Stanley Sawyer, John H. McDonald, Andrew Clark, Wen-Hsiung Li and Yun-Xin Fu, Brian Golding, and Richard Hudson provide different theoretical treatments of the consequences of natural selection. Even if most genetic variation is neutral, there would be a strong interest in finding those few that are influenced by selection. These papers show different consequences of the actions of natural selection and illustrate how one might use these to search for evidence of and/or to measure selection in real populations. Sawyer develops tests to estimate selection and shows that even synonymous nucleotide sites are subject to natural selection. McDonald makes use of geographic genetic differences to detect selection. Clark extends the application of these methods to genetic characters with a continuous distribution. Li and Fu review the methods available to measure population parameters. Golding looks at maximum likelihood methods to measure selection coefficients from species phylogenies, and Hudson looks at some consequences of deleterious, recurrent mutations.

The papers by Robert DeSalle and Alfried Volger; Patrick Dennis; Andy Beckenbach; and Donal Hickey, Shaojiu Wang, and Charalambos Magoulas examine some of the consequences of these types of data. DeSalle and Volger look at whether or not cladistic methods can be applied to intraspecific problems and what would be necessary to make this viable/useful. Dennis examines the selection driven divergence of genes adapting to high salt environments. Beckenbach shows that despite any potential limitations of the neutral theory, it can still explain even very unusual patterns of molecular divergence. Hickey et al. examine the molecular evolution of amylase genes within different insect species. They also demonstrate that deviations from simple neutral models do not necessarily implicate selection.

The papers by Jody Hey and Richard Kliman, Rama Singh and Ling-Wen Zeng, and Naoyuki Takahata deal with the events of speciation. Hey and Kliman show that the same sequence data used to explore the effects of selection can also be informative about the passage of genetic variation through a speciation event. Singh and Zeng demonstrate that molecular divergence and the events of speciation may be uncoupled. Takahata shows that there is an optimal level of genetic variation at the *Mhc* loci and shows the novel phenomena that this may contribute to the genetic isolation that separates populations and species.

The Canadian Institute for Advanced Research (CIAR) is a private, non-profit corporation that has mandated itself to aid interaction and networking of scientists in different areas. This is not an easy challenge given the current level of financial restraint, but the CIAR continues to promote excellence within the scientific community. The CIAR is proud to have sponsored this workshop to bring these aspects of evolution studies into focus and to bring both experimentalists and theoreticians together to look at this basic problem in biology.

BRIAN GOLDING
May 1994

Contributors

Monserrat Aguadé, Department de Genetica, Facultat de Biologia, Universitat de Barcelona, Spain.

Charles F. Aquadro, Section of Genetics and Development, Biotechnology Building, Cornell University, Ithaca, NY 14853.

Andrew T. Beckenbach, Institute of Molecular Biology and Biochemistry, Department of Biological Sciences, Simon Fraser University, BC, Canada.

David J. Begun, Section of Genetics and Development, Biotechnology Building, Cornell University, Ithaca, NY 14853.

Andrew G. Clark, Department of Biology, Pennsylvania State University, University Park, PA 16802.

Patrick P. Dennis, Department of Biochemistry and Molecular Biology, University of British Columbia, Vancouver, BC, Canada.

Rob DeSalle, Department of Entomology, American Museum of Natural History, New York, NY 10024.

Walter F. Eanes, Department of Ecology and Evolution, State University of New York, Stony Brook, NY 11794.

Yun-Xin Fu, Center for Demographic and Population Genetics, University of Texas, Houston, TX 77225.

John H. Gillespie, Center for Population Biology, University of California, Davis, CA 95616.

Brian Golding, Department of Biology, McMaster University, Hamilton, Ontario, Canada.

Jody Hey, Department of Biological Sciences, Rutgers University, Piscataway, NJ 08855.

Donal A. Hickey, Department of Biology, University of Ottawa, Ottawa, Canada.

Richard R. Hudson, Department of Ecology and Evolutionary Biology, University of California, CA 92717.

Norman L. Kaplan, National Inst. of Environmental Health Sciences, P.O. Box 12233, Research Triangle Park, NC 27709.

Eric C. Kindahl, Section of Genetics and Development, Biotechnology Building, Cornell University, Ithaca, NY 14853.

Richard M. Kliman, Department of Biological Sciences, Rutgers University, Piscataway, NJ 08855.

Charles H. Langley, Center for Population Biology and the Section of Evolution and Ecology, University of California, Davis, CA 95616.

Wen-Hsiung Li, Center for Demographic and Population Genetics, University of Texas, Houston, TX 77225.

Charalambos Magoulas, Department of Biology, University of Ottawa, Ottawa, Canada.

John H. McDonald, Department of Biology, University of Delaware, Newark, DE 19710.

Stanley A. Sawyer, Department of Mathematics, Washington University, St. Louis, MO 63130.

Stephen W. Schaeffer, Department of Biology, The Pennsylvania State University, University Park, PA 16802.

Rama S. Singh, Department of Biology, McMaster University, Ontario, Canada.

Wolfgang Stephan, Department of Zoology, University of Maryland, College Park, MD 20742.

Naoyuki Takahata, Department of Genetics, The Graduate University of Advanced Studies, Mishima, Japan.

Alfried P. Vogler, Department of Entomology, American Museum of Natural History, New York, NY 10024.

Shaojiu Wang, Department of Biology, University of Ottawa, Ottawa, Canada.

Ling-Wen Zeng, Department of Biology, McMaster University, Ontario, Canada.

1

Alternatives to the Neutral Theory

John H. Gillespie

Introduction

In the past few years, there has been increasing evidence that much of molecular evolution may not involve neutral alleles. This is almost certainly the case from proteins[5] and may be true for silent sites in coding regions as well.[9,14] Should we feel compelled to reject the neutral theory for certain classes of molecular variation, there is a conspicuous vacuum in the literature when we seek alternative models. The time seems ripe, therefore, to begin an exploration of non-neutral models to see how compatible they are with the very observations that cast doubt on the neutral theory.

I have written a computer simulation program that examines molecular evolution and variation in six population genetic models that have, at one time or another, been proposed as models of molecular evolution. They are the neutral, overdominance, underdominance, house of cards, SAS-CFF and TIM models. The simulations have uncovered a wealth of interesting dynamics that provide fodder for the theoretician, but little solace for the experimentalist.

One important implication of the simulations concerns the common observation that rates of protein evolution are variable. We have known for some time that the variance in the number of amino acid substitutions is from 2.5 to 7.5 times that expected under the neutral model.[5] The simulations indicate that rate variation this high is not easily accounted for by any of the models examined. The only reasonable alternative appears to be a new class of models that include parameters that change on very long time scales.

The Simulation Program

The simulation program contains two main components: the first deals with allele frequency dynamics, the second with the model of the gene. The two are virtually independent, so changes may be made in the allele frequency dynamics without altering that part of the program concerned with the genealogies.

The allele frequency component models a simple Markov process of variable dimensionality. At the beginning of each generation, there are K alleles in the population with frequencies $x_1, x_2, \ldots, x_k$. The frequencies are first changed deterministically by whichever model of selection is in vogue. Next, multinominal sampling occurs which will, in general, change both the frequencies of the alleles and their total number, K. Finally, mutation to new alleles occurs. This, too, will change K.

The selection models are as follows:

Neutral model For this simplest of all selection models, the allele frequencies are left unchanged after the selection step.

Overdominance model The symmetrical form of the overdominance model is used. The model requires a simple parameter, $\alpha = 2N\sigma$, where σ is the selection coefficient against homozygotes and N is the population size. The stationary properties of the diffusion approximation of this model are well known.[22,21]

Underdominance model The symmetrical underdominance model is exactly like the overdominance model except that heterozygotes are at a selective disadvantage σ when compared to homozygotes.

SAS-CFF model The symmetrical SAS-CFF model is a model of selection in a random environment.[5] The model requires two parameters: $\alpha = 2N\sigma$, where σ is now the standard deviation of the environmental contribution to fitness, and $B > 1$, which is a measure of the strength of balancing selection relative to that of disruptive selection.

TIM model The TIM model[18,19] is formally a model of haploid selection in a temporally fluctuating environment but also approximates an additive diploid model in a slowly changing environment. It is the same as the SAS-CFF model with $B = 1$. There is no balancing component of selection in the TIM model.

House of cards model This is an additive model where fitnesses are assigned at random to each new mutation and are never changed.[11] It is frequently used as a model of mildly deleterious molecular evolution as the population will quickly evolve to an exalted state where the vast majority of new mutations are harmful.[12,15]

The model of the gene used in the simulation is the infinite-sites, no recombination model.[20] This model defines an allele genealogy where each node represents a new mutation at a site. In the computer simulation, each node of the genealogy also records the abundance of the node and its time of origin and, should it fix, the time of fixation.

The simulation was written in C++, a language that is particularly well-suited for molecular evolution simulations, and were run on Silicon Graphics Indigos.

Substitution Processes

There are two substitution processes: The *origination process* is the point process of times at which mutations at sites that ultimately fix in the population first appear. The *fixation process* is the point process of times at which mutations at sites fix in the population. Figure 1 illustrates the connection between the two. The figure gives the correct impression that the origination process is relatively simple, with widely spaced events, while the fixation process is complex, with multiple fixations occurring in the same generation. Fortunately, the origination process is the only one that can be investigated experimentally.

Associated with each substitution process is a counting process, $\mathcal{N}(t)$, that gives the total number of events in an interval of t generations. The counting process is particularly important for evolutionists as their data consists of the total number of differences between a pair of species.

In the following we will investigate two aspects of substitution processes, the rate of substitution and the second order moments of the times between substitutions.

The rate of substitution is defined as

$$\mathrm{k} = \lim_{t\to\infty} \frac{\mathcal{N}(t)}{t} = \frac{E\mathcal{N}(t)}{t},$$

where $\mathcal{N}(t)$ may refer to either the origination or substitution process. Figure 2 illustrates the rate divided by the mutation rate as a function of α. (Recall that under the neutral model the rate of substitution is equal to the mutation rate.)

Figure 2 shows that the models fall naturally into two groups. For the overdominance, SAS-CFF, and TIM models, the rate of substitution increases with the strength of selection, while for the underdominance and house of cards models, the rate decreases with α. The dichotomy is easily understood; for the former

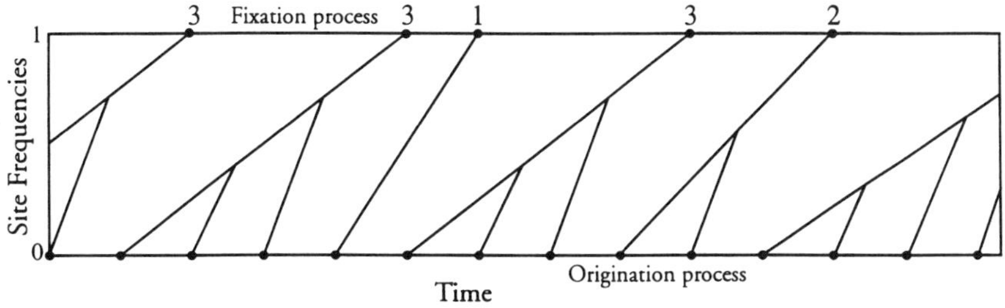

Figure 1. A diagram of the trajectories of mutations that ultimately fix in the population.

group, rare alleles are at a selective advantage, while for the latter group, they are at a selective disadvantage. It is perhaps surprising how quickly the rate drops off with α for the underdominance and house of cards model. An explanation will be given after the discussion of the first group of models.

Figure 2 shows that the rate of substitution is a concave function of alpha for the overdominance, SAS-CFF, and TIM models. The concavity is not entirely expected for those steeped in the standard arguments against selection as a cause of molecular evolution. For example, Kimura[10] repeatedly uses the expression

$$k = 4N_e s\upsilon$$

as the rate of substitution under a mutation-limited directional selection model that is only vaguely defined. From this, it is tempting to conclude that the rate of substitution is a linearly increasing function of both the strength of selection, s, and the mutation rate, υ. Figure 2 shows that this is not, in general, the case.

The fact that the concavity increases as the models move through TIM, SAS-CFF with $B = 1.5$, overdominance, and SAS-CFF with $B = 5.0$ gives a clue as to its cause. The order goes from a model with no balancing component to its selection, the TIM model, through models with increasing balancing selection components. As heterozygosity increases with increasing balancing selection, it is natural to expect that the concavity is due to the allelic diversity in the population. While heterozygosity does contribute, the complete explanation is more complicated.

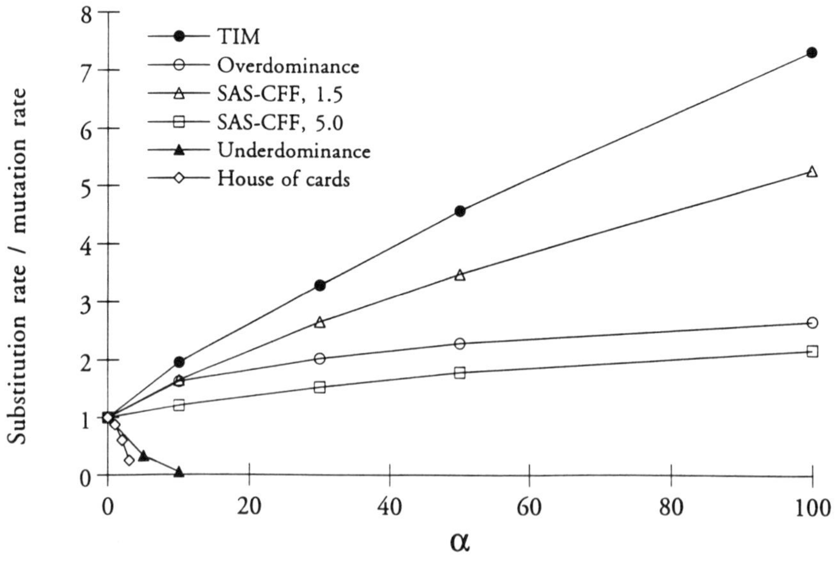

Figure 2. The rate of substitution as a function of the strength of selection, α, for five models with $\theta = 1.0$ and $N = 1000$.

TIM selection has no balancing component. As the strength of selection increases ($\alpha \rightarrow \infty$), the heterozygosity decreases and the rate of substitution increases. For sufficiently strong selection, a two-allele approximation to the dynamics gives[6,19]

$$k \sim u\alpha/\log(\alpha). \tag{1}$$

The division by $\log(\alpha)$ yields the concave dependency of k on α. However, as $\log(\alpha)$ changes very slowly with α, the degree of concavity is slight, certainly not nearly as extreme as it is for the overdominance model.

The reason for the extreme concavity for models with strong balancing components of selection may be understood by noting that Equation 1 was derived by the standard argument

$$k = 2Nu \times \text{‘fixation probability of a new allele,’} \tag{2}$$

where the fixation probability comes from a two-allele calculation. While this is fairly accurate for the TIM model, it breaks down entirely for the other models. For example, for the overdominance model with $\alpha = 50$, equation 2 yields $k/u = 47.75$, which is about 21 times the rate given in Figure 2.[7]

The problem arises because we confused the fixation of alleles and the fixation of sites. For neutral models and two-allele models, this distinction is unimportant for calculation of rates. However, once balancing selection becomes important, the ultimate fate of a mutation at a site depends on the allele that the first mutation appeared in and all of the alleles descended from it. Unlike neutral alleles, a collection of selected alleles do not have the same dynamics as a single allele with frequency equal to the sum of the alleles in the collection. Thus, the probability of fixation of a site will be extraordinarily difficult to calculate in general. (There are asymptotic methods available for strong selection and weak mutation,[5] but we are not concerned with such parameters here.)

A simple argument yields a quick approximation that is a vast improvement over equation 2. The fixation of a mutation at a site is a two-step process. The first step is the escape of the mutation from the boundary layer, the second is the *genealogy drift* of the mutation through the segregating alleles. For symmetrical models, the genealogy drift of a mutation through K segregating alleles is similar to the drift of a mutation through a haploid neutral population of K individuals. Thus, the probability of genealogy fixation of an escaped mutation is approximately $1/K$.

The probability of escape of a rare allele is usually easy to calculate. This probability times the mutational input each generation ($2Nu$) times $1/K$ gives the rate of substitution. For the overdominance model it is[7]

$$\frac{4Nu\sigma\, E\mathcal{F}}{\mathrm{EK}}. \tag{3}$$

Using values of $E\mathscr{F}$ and EK from the simulation, equation 3 gives $k/u = 1.422$, compared to 2.27 from the simulations. The relative error is 37%, which is a tremendous improvement over the 1994% relative error that comes from using formula 2. Thus, the purposefully vague formula,

$$k = 2Nu \times \text{`escape probability'} \times \frac{1}{K}$$

more faithfully captures the rate of substitution of sites than does the venerable equation 2.

The description of the rate as a two step process shows that the concavity comes from two sources. As the heterozygosity increases, it is harder for alleles to escape from the boundary layer. Should they escape, it becomes harder for them to drift through the genealogy to fixation. As the balancing component of selection increases, the latter factor becomes more important. In the case of overdominance and SAS-CFF with $B = 5$, genealogy drift becomes a major deterrent to the fixation of sites. As a consequence, the rate of substitution becomes remarkably insensitive to the strength of selection. Figure 2 shows that the rate varies only about two-fold as α varies from zero to one fifty for the overdominance and SAS-CFF with $B = 5$ models. Insensitivity to parameters has been the keystone argument for the neutral theory. Our observations blunt the force of the argument.

The next problem posed by Figure 2 concerns the steep decrease of the rate of substitution with alpha for the underdominance and house of cards models. The explanation for the house of cards model is particularly interesting.

Anyone who has tried to simulate molecular evolution with the house of cards model has had the experience of waiting an intolerable length of time for the simulations to complete when $\alpha > 3$. It seems that evolution comes to a dead standstill. In fact, for $\alpha > 5$, it does for all intents and purposes. The reason is tied up with the theory of records.

Consider the following problem. Draw a random variable from some probability distribution and note its value. Next, draw additional random variables from the same distribution until one exceeds in value the original random variable. On average, how many draws are required when this experiment is repeated many times? The surprising answer is that the mean number of draws is infinite. The proof of this classic paradox is probability theory may be found in any standard text, for example, on page 15 of Feller.[1] Feller places this result in a section titled "The Persistence of Bad Luck."

The house of cards model in large populations has persistent good luck. When a simulation is begun, a fitness is drawn at random and assigned to the first allele. A substitution occurs when a new mutant appears with a fitness that exceeds that of the original allele. Thus, the expected time to completion of the stimulation is infinite!

Of course, the usual situation in any particular simulation is for several substitutions to occur before stagnation. The theory of records gives some insight into the number of substitutions that are likely to occur. Returning to the original probability problem, we will call a particular draw a record if the value of the random variable exceeds that of the maximum value of all preceeding draws. The mean number of records in n draws is[8]

$$ER_n = \sum_{i=1}^{n} 1/i \sim .5772 + \ln(n).$$

The mean number of records in ten draws is 2.93, for 100 draws it is 5.19, and for a million draws it is only 14.39. A million draws after the first million draws will raise the total mean number of records by only 0.69 records. The number of substitutions will be much less than this since the probability that a new advantageous mutation (a record) will enter the population is only twice its selective advantage.

For any particular simulation, the times between successive substitutions increases. Thus, the first will occur relatively quickly, the second will take longer, and so forth. In evolutionary terms, with each substitution the fitness of the fixed allele is raised, making the wait for the next advantageous mutation progressively longer.

Given these observations, why is the rate of substitution in Figure 2 not zero? In finite populations, it is possible for a less fit mutation to replace a more fit allele. In order for this to happen, selection must be very weak. Empirically, we find that α must be less than three or four to prevent stagnation, verifying Tachida's result.[15]

Ironically, for such small values of α, the population oscillates between substitutions of favorable and deleterious mutations. Therefore, the house of cards model is not a model of the evolution of deleterious mutations as is often claimed. More details on the house of cards model will appear in the third paper of my 'Substitution processes in molecular evolution' series.

Consider next the covariance structure of substitution processes. Substitution processes are stationary point processes. The times between originations or fixations form a stationary time series in discrete time. As time series are much easier to analyze than point processes, they will be the subject of this section.

Call the times between originations or fixations . . . , T_{-1}, T_0, T_1, . . . and define

$$C_k = \frac{c_k}{(ET_i)^2},$$

where

$$c_k = \mathrm{Cov}(T_i, T_{i+k})$$

is the covariance of two intervals separated by a lag of k originations, fixations, or bursts. The function C_k is closely connected to the index of dispersion of the counting process,

$$I(t) = \frac{\mathrm{Var}\mathcal{N}(t)}{E\mathcal{N}(t)}.$$

For large t,

$$R = \lim_{t\to\infty} I(t) \tag{4}$$

$$= C_0 + 2\sum_{i=1}^{\infty} C_i. \tag{5}$$

R is the quantity that is widely used to describe the variability in rates of substitution.[10] For proteins, $R \approx 7.5$.[5]

Representative values of C_i for three origination processes are illustrated in Figure 3. The figure shows that the neutral model has $C_0 = 1$ and $C_i = 0$, $i >$

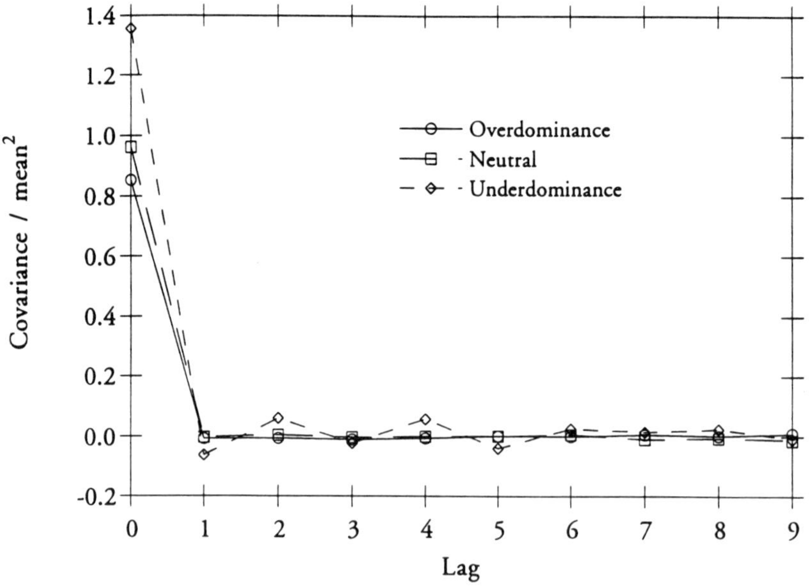

Figure 3. The covariance of the times between originations divided by the square of the mean, C_k. ($\theta = 1$, $N = 1000$, and $\alpha = 10$ for the overdominance model and $\alpha = 5$ for the underdominance model.)

1 as expected. Recall that the origination process for the neutral model is a Poisson process and that for the Poisson process the times between successive events are independent (hence $C_i = 0, i > 1$) and exponentially distributed (hence $C_0 = 1$).

The covariance structure for the overdominance origination process is a surprise. $C_0 < 1$ suggesting that the originations are more regularly spaced than they are for the Poisson process. $C_i \approx 0, i > 1$ suggesting that successive intervals are uncorrelated and thus that the origination process is indistinguishable from a renewal process. This same pattern is seen for the TIM and SAS-CFF models but not for the underdominance model.

By contrast, the fixation processes of these same models are more clustered than a Poisson process as suggested by Figure 1 and documented elsewhere.[7] We have an interesting dilemma. Why, for example, are the fixations of the overdominance model more clustered than those of the neutral model while the overdominance originations are more regular? It seems as if the covariance structure of the origination and fixation processes have little to do with each other. But this appears nonsensical given that an origination is identified by the fixation of the site.

Our intuition is helped immeasurably by a remarkable characterization of the origination process of the neutral model due to Sawyer.[13] He pointed out that the origination process can be defined in terms of the allele frequency process without any reference to fixations or genealogies. We will call his point process the Sawyer point process rather than the origination process as it does not correspond to the origination process for all models.

The Sawyer process is defined as follows. Consider a stationary, infinite-allele process. Choose an allele at random weighted by its frequency. (That is, the probability of choosing a particular allele is just its frequency.) Call this allele the *allele of record*. Follow the allele of record until it is lost from the population. Record the time of its loss and at that time choose a new allele of record at random weighted by its frequency. Follow the new allele of record until it is lost, record that time, choose a new allele of record, and so forth. The point process made up of the times of loss of alleles of record as described is not stationary as the first drawing it not at the moment of loss of an allele of record. Eventually it will achieve stationarity. The stationary point process is the Sawyer point process.

Sawyer argued that the Sawyer process (he didn't call it the Sawyer process) and the origination process are the same for the neutral model. The only aspect of the neutral model used by Sawyer was reversibility and thus that the Sawyer process and the origination process are the same for any reversible allele frequency model. It is known that the symmetrical overdominance and underdominance models are reversible, but not the TIM nor SAS-CFF models.[7]

The Sawyer process shows that the time between originations is just the time until a newly chosen allele of record is lost from the population. Thus, the essence of the origination process is a waiting time distribution. Waiting time properties of diffusion processes are much better understood than are the properties of allele genealogies and we are in a better position to understand the regularity of originations under selection models. While better, our position still isn't very good as it is quite difficult to carry out actual calculations for infinite-dimensional diffusions.

An asymptotic approach based on small mutation rates is tractable as there are usually only one or two alleles in the population. In this case, two-allele theory makes the appropriate calculations possible. A two-allele diffusion with reversible mutation matches very closely an infinite-allele diffusion as $\theta \rightarrow 0$. If the two-allele population is started with one allele fixed, the Sawyer process is initiated by making the allele of record the fixed allele. The first event occurs when the frequency of the allele of record hits zero, at which time the new allele of record becomes the allele with frequency one. Clearly, the times between events form a renewal process and have the same distribution as the time to fixation of an allele with initial frequency zero.

Let T be the waiting time for a one-dimensional process that begins at zero to hit one for the first time. The Sawyer process for this diffusion will be more regular than random if

$$C_0 = \frac{\mathrm{Var}T}{(ET)^2} < 0.$$

This inequality has been proven for any diffusion with reflecting barriers.[6] Thus, the regularity of the origination process should not cause us any concern, at least for small θ.

But what about the neutral and underdominance models? Why are they not more regular than a Poisson process? The answer illustrates a limitation in the two-allele approach. As $\theta \rightarrow 0$, $C_0 \rightarrow 1$ for all models, both infinite-allele and two-allele. For the neutral model, the infinite-allele C_0 equals one and the two-allele C_0 approaches one from below. For the overdominance, TIM, and SAS-CFF models, the infinite-allele and the two-allele C_0 both approach one from below, but they approach each other faster than either approaches one. Thus, the two-allele version give a very good approximation to the infinite-allele version for these three models.

For the infinite-allele underdominance model, C_0 approaches one from above while for the two-allele version it approaches one from below. The aspect of the underdominance model that leads to the clustering of originations is that the entry of alleles into the population is easier when the homozygosity is lower as may be seen in drift coefficient, $\sigma x_i(x_i - \mathcal{F})$. When one allele is moving toward fixation, it lowers $\mathcal{F}$ and makes it easier for a second allele to enter. As this synergism requires three alleles, it cannot be present in the two-allele version of the underdominance model.

The Sawyer process can also help us to understand why the origination process for the overdominance, TIM, and SAS-CFF and neutral models are so similar to a renewal process. For these, the alleles of record persist in the population long enough for their dynamics to essentially forget their initial frequencies. Thus, the long time of loss of an allele of record is not an improbable moment in the infinite-allele dynamics. When a new allele of record is chosen it is as if it were being chosen at a time that is independent of the history of the process. In other words, the process almost starts fresh with each new allele of record, which gives the renewal-like behavior. By contrast, underdominant or deleterious alleles are lost quickly, making the moment of their loss an improbable moment in the infinite-allele dynamics. As can be seen in Figure 3, the underdominance model does not mimic a renewal process.

I have examined many simulations for evidence that the origination processes for the overdominance, TIM, and SAS-CFF models deviate significantly from a renewal process. As yet, I have been unable to find parameter values that give a significant departure. It may be—although it seems unlikely—that they are renewal processes. A proper mathematical investigation would be a difficult but worthwhile endeavor.

The origination process for the house of cards model is radically different from the others, as illustrated in Figure 4. As the strength of selection increases, the clustering of substitutions becomes extreme, with $C_0 = 84$ when $\alpha = 3.5$. The reason is, once again, tied up with the theory of records. Should a particularly

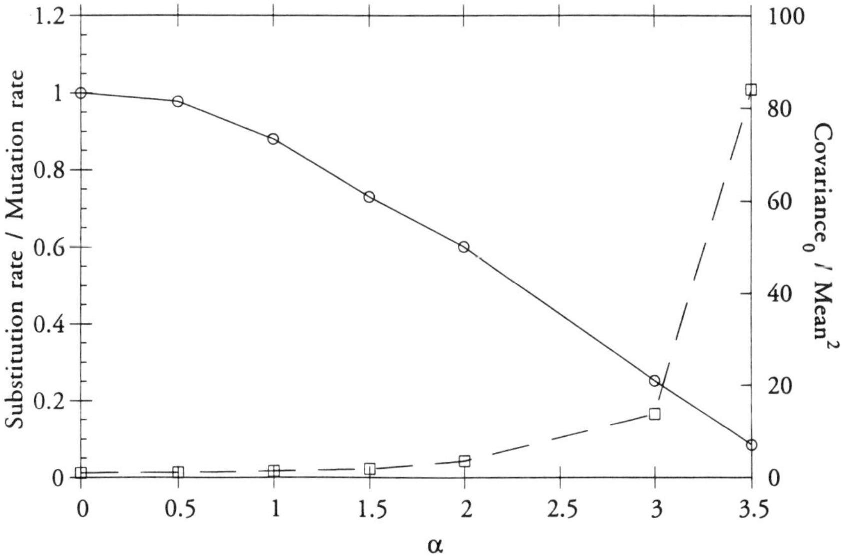

Figure 4. The rate of substitution over the mutation rate and the covariance over the mean squared for lag zero for the house of cards model. ($\theta = 1$ and $N = 1000$.)

good allele fix in the population, the waiting time until it is displaced by an even better allele can be very long (the mean time is infinite). More often than not, a deleterious allele will fix first. When this happens, the deleterious allele is relatively quickly replaced by an advantageous allele. Thus, the variation in the time between fixations is very large relative to the square of the mean. A complete mathematical treatment of the house of cards model will appear elsewhere.

Variation Within Populations

In contrast to the surprising results on substitution processes, the results on variation within populations are pretty much as expected. Figure 5 illustrates the dependency of the average sum of site heterozygosities on alpha for five models. All simulations use $\theta = 1$ for which the mean sum of site heterozygosities for the neutral model is one. Overdominance leads to the greatest diversity; house of cards to the least. Interestingly, both SAS-CFF models exhibit levels of variation that are not very different from the neutral model. While the SAS-CFF model does have a balancing selection component, it does not easily maintain high levels of variation.

A note of caution is in order. While the overdominance model always increases variation and the TIM, underdominance, and house of cards models always decrease variation when compared to a neutral model with the same θ, the situation with SAS-CFF models is more complicated. For a given θ, the site heterozygosity is a complicated, and largely unknown, function of both B and α. For a low value of B and a high value of θ, the heterozygosity may decrease with increasing α. For a high value of B, the heterozygosity may increase with α. In this sense, Figure 5 is misleading for SAS-CFF models.

Perhaps the most useful information to be gleaned from these simulations concerns hypothesis testing. There are several published procedures designed to test for agreement with the neutral model using samples from a single population. The most popular of these is Tajima's[16] test which is based on the statistic

$$D = \frac{\hat{k} - S/a_1}{\sqrt{e_1 S + e_2 S(S-1)}},$$

where $\hat{k}$ is the average site heterozygosity and S is the number of segregating sites in the sample. The other symbols are constants that depend on the sample size. The constants are chosen such that the mean of D is approximately zero and the variance is approximately one. Tajima has shown via simulations that the distribution of D is not normal, but is at least bell-shaped for large samples.

Figure 6 illustrates the behavior of D when the entire population is sampled. The shaded region is the neutral expectation plus and minus one standard deviation. As is obvious, Tajima's statistic is not capable of distinguishing between these

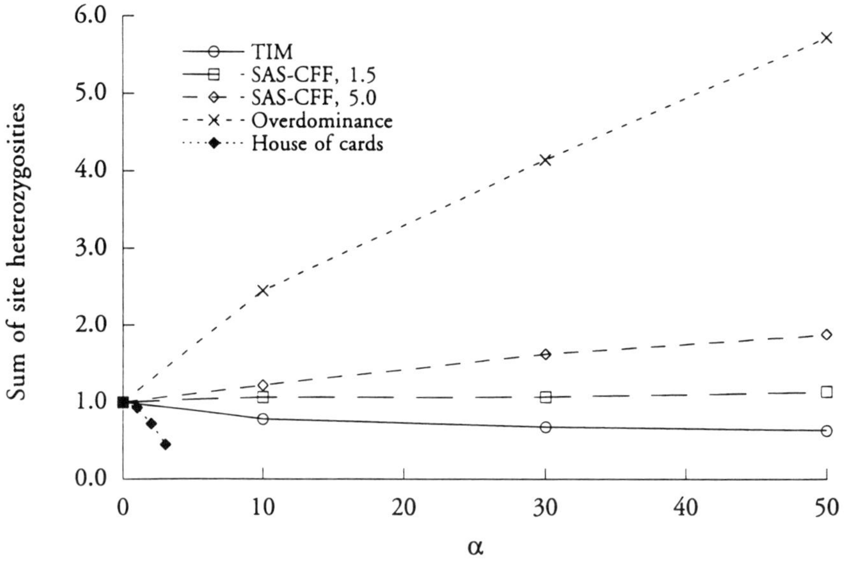

Figure 5. The average sum of site heterozygosities as a function of the strength of selection. ($\theta = 1$ and $N = 1000$.)

models except for very strong overdominance. The TIM and SAS-CFF models, in particular, are very similar to the neutral model.

Tajima's statistic is based on a comparison of estimations of θ based on the site heterozygosity and the number of segregating sites. A positive value indicates that the site heterozygosity is high relative to the number of segregating sites. The simulations show that $D > 0$ for those models with balancing components to their selection—overdominance and SAS-CFF—as expected. The TIM and house of cards models, on the other hand, exhibit negative values for D.

Once again, a cautionary note about SAS-CFF models is in order. Tajima's D under the SAS-CFF model exhibits more complicated behavior than suggested in Figure 6. Negative values are possible for small values of B. A complete characterization of D under SAS-CFF models will make an interesting study.

It is often assumed that most of the observed molecular variation is neutral and relatively little is selected. In regions of low recombination, the presence of selected sites will distort the variation at neutral sites. The reduction in variation seen in regions of low recombination in *Drosophila* is frequently attributed to linked hitchhiking events. However, as reported at this meeting, the molecular variation that is seen in these regions does not exhibit the large negative values of D predicted by models with simple selected sweeps. John Braverman (pers com) has found that a ten-fold reduction in neutral variation should be accompanied by a mean D of about -2.2. Such low values are not seen in the data reported by Aguade and Langley at the meeting.

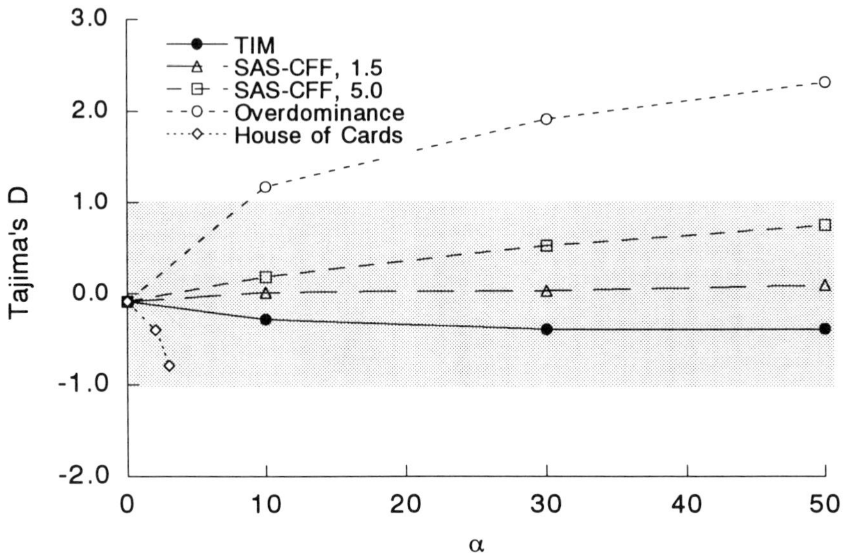

Figure 6. Tajima's D statistic as a function of the strength of selection. ($\theta = 1$ and $N = 1000$.)

As the TIM model will reduce variation without a significant change in D, it might be expected that it will have the same effect on linked neutral sites. Figure 7 shows that this is, indeed, the case. When the site heterozygosity of neutral sites is reduced to about one-tenth its neutral value, mean D is about -1.0, well within the neutral confidence limits.

Linked TIM selection is considerably more complex than linked selective sweeps. TIM alleles typically move around rapidly to ultimate fixation or loss. Unlike directional selection, TIM selection often results in alleles attaining moderate frequencies followed by loss. It should not surprise us that the correlation between D and heterozygosity for neutral sites for linked TIM selection is very different than for linked selective sweeps.

As the TIM model may be viewed as a limiting case of the SAS-CFF model as $B \rightarrow 1$, it is expected that neutral sites linked to SAS-CFF sites behave like those linked to TIM sites when B is close to one. Figure 7 illustrates this point.

The suggestion that selective sweeps are responsible for reduced variation in regions of low recombination is perhaps naive. It is my conviction that most selection is relatively weak (although $\alpha \gg 1$). In addition, it is my conviction that the environmental factors that changed to make an allele advantageous are likely to change again, and again . . . . For many mutations, their advantage will not last long enough for them to fix in the population. Thus, selected mutations will generally exhibit the sort of dynamics typified by TIM and SAS-CFF models. Consequently, neutral mutations will be affected by this selective

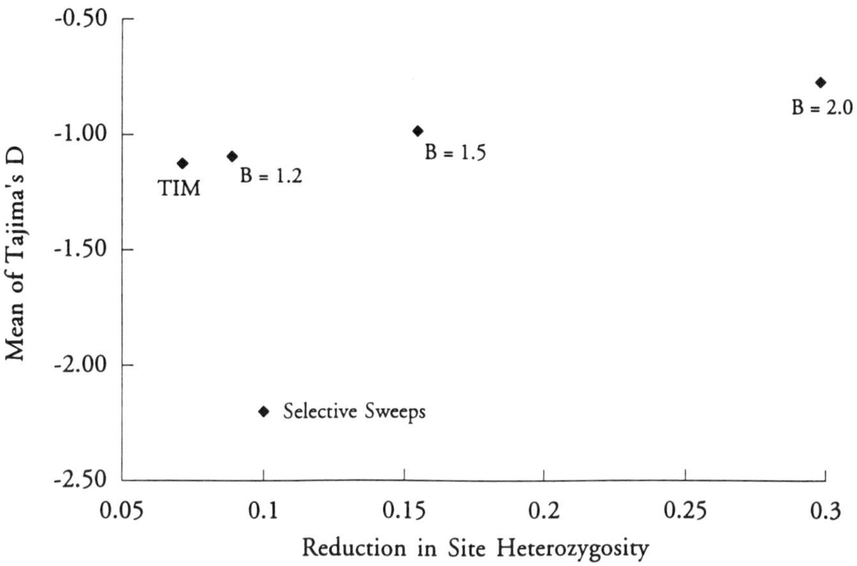

Figure 7. Tajima's D statistic as a function of the site heterozygosity for neutral mutations embedded in regions of TIM selection with no recombination. (Selected $\theta = 1.0$, $\alpha = 500$, neutral $\theta = 1.0$, and $N = 5000$.)

milieu more often than by selective sweeps. The observation that D for regions of low recombination and reduced variation is not large and negative supports this view.

Discussion

The results on genetic variation within populations are quite preliminary, but do suggest that it may be difficult to find evidence for selection even when the strength of selection is quite strong. In particular, the dynamics of the TIM and SAS-CFF models are similar enough to those of the neutral model to make hypothesis testing problematic. Most of the test—like the Tajima test—for neutrality have insufficient power to reject TIM and SAS-CFF models. We are really in the same position as we were in the 'electrophoretic era' when it was discovered that neutral alleles and SAS-CFF alleles have the identical sampling distributions.[2]

More compelling—in a negative way—is the failure to find any model that will easily account for the large variance in rates of substitution seen in protein evolution. The models with balancing components to selection actually give rates of substitution that are more regular than Poisson. The underdominance model has a slight tendency to give clustered substitutions, but the effect is not nearly large enough to account for a variance in substitutions that is seven times the mean. Only the house of cards model exhibits a large variance in $\mathcal{N}(1)$. However, the extreme dependence of this model on the parameter α makes it a poor candidate. If $\alpha < 1$, the model behaves just like the neutral model and can't account for the high variance of substitutions. If $\alpha > 4$, molecular evolution comes to a screeching halt. Thus the model is only compatible if $1 < \alpha < 4$, a condition that is too restrictive to take seriously.

The only satisfying resolution appears to be that some parameter changes on a time scale that is similar to that of molecular evolution. Gillespie[3,5] has suggested that fitness may fluctuate on this time scale (as well as other time scales) and Takahata[17] has suggested that the mutation rate may change on this time scale. The suggestion that the population size changes on such a long time scale has generally been discounted as a plausible explanation.[4,17] That a fluctuating mutation rate is the cause requires some special additional assumptions as the rate of silent substitution appears not to vary to the same extent as does amino acid substitution. Takahata included these assumptions in a model he called the 'fluctuating neutral space model'.

A corollary of our conclusion is that we must reject the standard models of population genetics (those models investigated here) and replace them with a new set that incorporates long time-scale changes in parameters.

References

1. W. Feller. (1966). *An Introduction to Probability Theory and Its Applications,* vol. II. John Wiley & Sons, Inc., New York.
2. J. H. Gillespie. (1977). Sampling theory for alleles in a random environment. *Nature 266*, 443–445.
3. J. H. Gillespie. (1984). The molecular clock may be an episodic clock. *Proc. Natl. Acad. Sci. USA 81*, 8009–8013.
4. J. H. Gillespie. (1988). More on the overdispersed molecular clock. *Genetics 118*, 385–386.
5. J. H. Gillespie. (1991). *The Causes of Molecular Evolution*. Oxford Univ. Press, New York.
6. J. H. Gillespie. (1993). Substitution processes in molecular evolution. I. Uniform and clustered substitutions in a haploid model. *Genetics 134*, 971–981.
7. J. H. Gillespie. (1993). Substitution processes in molecular evolution. II. Exchangeable models from population genetics. *Evolution submitted*.
8. N. Glick. (1978). Breaking records and breaking boards. *Amer. Math. Monthly 85*, 2–26.
9. T. Ikemura. (1981). Correlation between the abundance of *Escherichia coli* transfer rnas and the occurrence of the respective codons in its protein genes. *J. Mol. Biol. 146*, 1–21.
10. M. Kimura. (1983). *The Neutral Allele Theory of Molecular Evolution*. Cambridge University Press, Cambridge.

11. J. F. C. Kingman. (1978). A simple model for the balance between selection and mutation. *J. Appl. Prob. 15,* 1–12.

12. T. Ohta. (1992). The nearly neutral theory of molecular evolution. *Annu. Rev. Ecol. Syst. 23*, 263–286.

13. S. Sawyer. (1977). On the past history of an allele now known to have frequency *p*. *J. Appl. Prob. 14*, 439–450.

14. D. C. Shields, P. M. Sharp, D. G. Higgens, and F. Wright. (1988). "silent" sites in *Drosophila* genes are not neutral: Evidence of selection among synonymous codons. *J. Mol. Biol. Evol. 5*, 704–716.

15. H. Tachida. (1991). A study on a nearly neutral mutation model in finite populations. *Genetics 128*, 183–192.

16. F. Tajima. (1989). Statistical method for testing the neutral mutation hypothesis by DNA polymorphism. *Genetics 123*, 585–595.

17. N. Takahata. (1987). On the overdispersed molecular clock. *Genetics 116*, 169–179.

18. N. Takahata, K. Iishi, and H. Matsuda. (1975). Effect of temporal fluctuation of selection coefficient on gene frequency in a population. *Proc. Natl. Acad. Sci. USA 72*, 4541–4545.

19. N. Takahata, and M. Kimura. (1979). Genetic variability maintained in a finite population under mutation and autocorrelated random fluctuation of selection intensity. *Proc. Natl. Acad. Sci. USA 76*, 5813–5817.

20. G. A. Watterson. (1975). On the number of segregating sites in genetic models without recombination. *Theor. Popul. Biol. 7*, 256–276.

21. G. A. Watterson. (1977). Heterosis or neutrality? *Genetics 85*, 789–814.

22. S. Wright. (1949). Adaptation and selection. In *Genetics, Paleontology, and Evolution,* G. L. Jepson, G. G. Simpson, and E. Mayr, Eds. Princeton Univ. Press, Princeton, pp. 365–389.

2

Patterns of Polymorphism and Between Species Divergence in the Enzymes of Central Metabolism

Walter F. Eanes

Abstract. *Recently, a number of the genes encoding the enzymes involved in the early steps of glycolysis and its branches have been cloned from* Drosophila melanogaster. *In several labs, elected enzymes have undergone a systematic analysis of the levels of nucleotide sequence polymorphism and interspecific divergence. Specifically in this lab, the enzyme for glucose-6-phosphate dehydrogenase shows a dramatic deviation from the neutral expectation when levels of synonymous polymorphism and divergence are contrasted with levels of amino acid polymorphism and divergence using* D. simulans *as a close relative. The analysis reflects a high rate of amino acid substitution relative to the level of amino acid polymorphism in the two species. The results for four other enzymes (GAPDH, GPDH, 6PGD, and PGI) either indicate little divergence in sequence, or patterns not significantly different from neutral expectation. The discord pattern for G6PD may arise because it regulates flux through the pentose shunt, making it a responsive target for natural selection. Preliminary data using* D. yakuba *G6PD sequence to establish the polarity of mutation events, suggests that the rates of amino acid and silent divergence are very different for the separate* D. melanogaster *and* D. simulans *lineages. Work is now underway in this lab to examine triosephosphate isomerase and aldolase as well.*

Introduction

Prior to the development of DNA methods, empirical population genetics emphasized the study of electrophoretic variation of soluble enzymes. Because methods for visually staining soluble enzymes generally incorporated cofactors (NADP, NAD, ATP), or coupled reactions to enzymes using cofactors, this inadvertently favored the screening of enzymatic steps in well-known pathways of central metabolism. For this reason we know that many of the early steps of glycolysis and its branches are polymorphic in *Drosophila*, as many other organisms. Figure 1, in a simplified fashion, shows the initial steps in glycolysis and branch points.

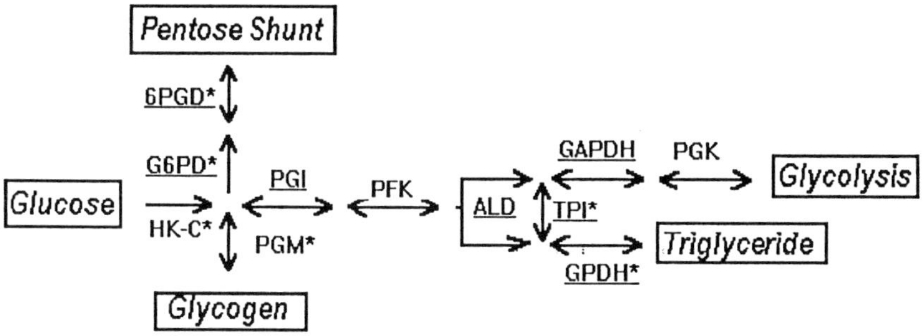

Figure 1. Generalized pathway scheme for the enzymes in the early steps of glycolysis and its branch points. The enzymes are phosphoglucomutase (PGM), hexokinase (HEX-C), glucose-6-phosphate dehydrogenase (6GPD), 6-phosphogluconate dehydrogenase (6PGD), phosphoglucose isomerase (PGI), phosphofructokinase (PFK), aldolase (ALD), triosephosphate isomerase (TPI), glycerol-3-phosphate dehydrogenase (GPDH), glyceraldehyde-3-phosphate dehydrogenase (GAPDH), and phosphoglucose kinase (PGK). The corresponding genes that have been cloned in *Drosophila melanogaster* are underlined and enzymes with common electrophoretic polymorphisms are marked with asterisks.

Enzymes possessing electrophoretic polymorphisms in *D. melanogaster* are marked with an asterisk, and those genes that have been cloned are underlined. Because of this early attention, a great deal is known of the geographic variation of electrophoretic variants of four these polymorphisms, notably glucose-6-phosphate dehydrogenase (G6PD), phosphoglucomutase (PGM), glycerol-3-phosphate dehydrogenase (GPDH), and 6-phosphogluconate dehydrogenase (6PGD), and triosephosphate isomerase (TPI). Others, such as hexokinase (HK, there are two loci)), have polymorphisms, but are not well characterized. It appears that four are electrophoretically monomorphic (PGI, ALD, PGK and GAPDH), although until recently we knew little about the possibility of electrophoretically silent amino acid polymorphisms for these enzymes. From a number of studies, some generalities have emerged. For instance, in many cases, there exist not only notable intercontinental differences in allele frequency, but also within continent latitudinal clines.[1-3] Along with ADH, and with the exception of 6PGD, there is a pattern of replacement of a "tropical" allele (or electromorph) by a "temperate" allele. The question is, can we interpret these polymorphisms as cases of adaptive change in these enzymes of metabolism, or do they simply represent biochemical noise? Are these old alleles, where the so-called temperate variant has persisted in a rare temperate (perhaps montane) niche in equatorial Africa, or do they represent new mutations, whose rapid rise is selection driven? Finally, are there points in the overall pathway where selection is most likely to get a response, as reflected by adaptive substitution, and does this bear any relationship to the propensity for amino acid polymorphism?

While often maligned as a "garbage species," *Drosophila melanogaster* has recently emerged as an evolutionary model, and along with its sibling species, *D. simulans* (and close relatives), a number of important population genetic questions are being addressed using a comparative approach. The prevailing view is that the global population of *D. melanogaster* represents several historical expansions,[4,5] and this global structure offers the opportunity to investigate selection operating over different time scales. There is increasing evidence for what must be a very old subdivision between east and west equatorial Africa. Differences between subSaharan Africa and Europe are also believed to reflect a strong historical subdivision. In contrast, the origin of New World and Australian populations are probably very recent (several hundred years), and have yet undetermined sources. Nevertheless, in these recently derived continental populations, latitudinal clines are pervasive for morphological and physiological phenotypes[4,6] (such as ethanol tolerance), as well as inversion and allozyme polymorphisms. The above enzymes that have been screened in any detail show clines associated with a latitudinal spread, perhaps seasonally from more moderate climates. Do these allozyme polymorphisms represent part of the genetic variation that contributes to potential in metabolic physiology? The proposal offered here is that a concerted study of the diversity and pattern of intra- and interspecific DNA sequence variation for these genes will illuminate the historical selection that may (or may not) be acting on these polymorphisms, and will also tell us about the propensity of various steps to contribute to adaptive change.

The Genealogical Approach

In 1972, with electrophoretic data in mind, Warren Ewens proposed what would be considered the first formal statistical test of the neutral model[7]. This test questioned (for a given sample size) whether the frequency profile for an observed number of alleles was expected under an infinite allele model of Kimura. Ewens immediately pointed out that the test was statistically weak; for a specific sample size and number of allele classes, many different frequency profiles are compatible with the same stochastic process, and natural selection is also capable of generating many similar frequency spectra. Gillespie has since shown that random selection can generate allele frequency profiles similar to neutral processes.[8] However, Ewens later emphasized that, if the genealogical relationships between all the genes in our sample were known, some competing hypotheses might be rejected[9]. Coalescence theory has recently been used to predict quantifiable features of genealogies under different forms of selection.[10,11] This would all be nothing more than an exercise in mathematical theory, if genealogies could not be somehow determined. The introduction of DNA sequencing, now greatly facilitated by PCR methods, allows the estimation of genealogies, as long as recombination does not obscure them.

The Studies of Adh, 6Pgd, Gpdh, Gapdh, and Pgi

While it is not one of the enzymes in the central pathway shown in Figure 1, the analysis of the *Adh* electrophoretic polymorphism has forged much of the path for this type of approach.[12-14] The fast, high activity, electrophoretic allele, long speculated to be a recent mutation selected to much higher frequencies as *Drosophila melanogaster* invaded temperate latitudes (ethanol tolerance increases as well), shows much less sequence diversity than the slow allele. The possibility that it is new, however, was dismissed by the observation that significant sequence divergence also exists between the two alleles. Therefore, the summary picture is one of an allele that has increased recently, but prior to this probably persisted at low frequency for a significant period of time. Recombination, while detectable in the sequence had not obscured the history of the polymorphism.

The case for glycerol-3-phosphate dehydrogenase (GPDH) is similar to that for ADH. Two electrophoretic alleles are found in the global population, and one allele increases as populations move away from the equator.[1] The slow (ancestral) allele predominates in equatorial Africa. Functionally, the polymorphism has been implicated in genetic variation for flight metabolism,[15] although the metabolic role must be diverse, given the gene is expressed in larvae and is differentially spliced.[16] Irrespective of the role, the polymorphism conforms to the above scenario, where an allele is rare in Africa, lower in sequence diversity, and becomes common in temperate regions.[17] The exons of the gene also carry high levels of silent polymorphism, and this is an expected feature of a region exposed to a site under balancing selection. However, recombination appears to be sufficient to have obscured much of the genealogical history of this polymorphism. The analyses of the rather rare electrophoretic polymorphism in *D. simulans* will be interesting. There are no replacement differences between *D. melanogaster* and *D. simulans,* but this should have been predicted because there is only one amino acid difference between *D. melanogaster* and *D. virilis.*[18] This is interesting because restricted divergence was predicted from earlier interspecific studies of electrophoretic variation.[19] Therefore, while there is some evidence for an adaptive amino acid polymorphism at GPDH, there is no evidence for selection driven amino acid substitution; it is a conservative enzyme in the pathway.

The case for 6PGD is one of a rare electrophoretic allele in Africa, and polymorphism in the New World and Australia, where the clines parallel G6PD (the low activity variants covary). Restriction map studies find the slow allele to possess almost no sequence variation (Begun and Aquadro, personal communication), although this allele has apparently recently increased in frequency to as much as 70% in North America. There is also no evidence for significant adaptive substitution in comparison of *D. melanogaster* and *D. simulans* DNA sequences using the McDonald-Kreitman approach. Studies are currently underway by John McDonald and Marty Kreitman to study PGI and GAPDH. Both enzymes lack electrophoretic polymorphisms in *D. melanogaster,* although PGI is generally a highly polymorphic enzyme in many species. Neither study finds any evidence for electrophoretically silent amino acid polymorphisms, and neither enzyme appears to be undergoing significant levels of amino acid replacement when analyzed against the levels of silent site divergence and polymorphism (Kreitman, personal communication).

The Case of G6PD in Drosophila melanogaster and Drosophila simulans

This lab has focused for a number of years on one of these enzymes, the X-linked glucose-6-phosphate dehydrogenase locus (*G6pd*). This is geographically the most variable of all allozyme polymorphisms in *D. melanogaster.*[2,3] The function of G6PD in *D. melanogaster* is well established as the initial enzymatic step in the pentose shunt pathway. The role of the pentose shunt is to maintain the NADP/NADPH balance of the cell in the face of demands that draw on the NADPH pool, such as fatty acid synthesis and detoxification. In *D. melanogaster*, it has been shown that 40% of the reduced NADPH is provided by this pathway.[20] Our investigations show that glucose flux through the pentose shunt is very sensitive to activity variation at G6PD, as might be expected for a regulatory enzyme located at the branch point between metabolic pathways.[21,22] Because of this important regulatory role at a branch point, G6PD might be expected to be relatively conservative and resistant to both amino acid polymorphism and fixation. Alternatively, if natural selection "favors" varying flux over time and space, then G6PD may well be the most responsive target. A suggestion of this emerges from the human G6PD, where much of the world's population, as a response to malarial selection, carry deficient G6PD variants of multiple origin; more directly stated, these individuals are deficient for red blood cell NADPH, and thus, their red cells offer a hostile environment for the malarial parasite.[23] While the selective pressure has been to lower NADPH in the red cell, the response has been via lowering G6PD activity. In our studies using DNA sequence data, we will be interested in examining these questions at both the intra- and interspecific levels.

As part of our studies,[24] we have sequenced 32 copies of the coding region of the *G6pd* locus in *D. melanogaster* and 12 copies in *D. simulans*, which has no electrophoretic polymorphism for G6PD. Using the logic applied in the study of ADH evolution in the *D. melanogaster-D. simulans-D. yakuba* lineages, we examined the concordance of levels of amino acid polymorphism and divergence with the same measures for silent or synonymous sites.

Table 1 summarizes the data for the 85 positions within the *G6pd* 1548 nucleotide coding region that are either polymorphic within, or different between the two species. The levels of silent polymorphism within each species are

Table 1. Summary data for number of segregating nucleotide polymorphisms in Drosophila melanogaster *and* D. simulans, *as well as the number of 'fixed' differences between the samples of the two species. The bottom row lists the number of site differences between two randomly selected lines (OK93 and DPF88S).*

	Replacement	Silent
Polymorphism		
D. melanogaster	2	22
D. simulans	0	14
Pooled	2	36
"Fixed"	21	26
Divergence *OK93 vs DPF88S*	21	35

similar, and no polymorphisms are shared. Assuming a Wright-Fisher model, the estimated neutral parameter Θ for exons 2-4 is 0.0036 and 0.0029 for the 32 *Drosophila melanogaster* and 12 *D. simulans* lines respectively. The corresponding pair wise differences (π) are 0.38 and 0.34 percent. These values are normal for these species. Our *D. melanogaster* sample shows two replacement site polymorphisms. These correspond to the well characterized *A/B* allozyme polymorphism distributed throughout the global population and involves a proline to leucine change at position 1817, and the unique *AF1* electrophoretic allele seen in European populations, that results from a glycine to cysteine change at nucleotide 619. *D. simulans* has no replacement polymorphisms in the sample of 12 copies, and no allozyme polymorphism has ever been reported or seen by this lab for this species.

All our observations indicate that this single proline to leucine change is the cause of the *in vitro* and *in vivo* differences we have described.[21] An earlier study, which screened for *in vivo* function 11 rare electrophoretic variants of G6PD, reported two classes of G6PD function, and we proposed that this clustering reflected the common allele of origin for each rare variant.[25] Our DNA sequences of these rare variants have confirmed this prediction, and further emphasize the importance of the Pro/Leu polymorphism in *in vivo* function. As also predicted, the temperate associated *A* allele is derived from the common tropical *B* variant, and possesses about a third the sequence diversity. The *AF1* allele shows no sequence diversity among multiple copies, and is essentially a common *A* haplotype with a single replacement change. This is interesting because this is a very common allele in many populations north of the Sahara.

We do not yet have all the data to carry out a series of HKA-like tests to examine the alternative hypothesis that the *Pro/Leu* polymorphism is older than expected under neutrality. The haplotype structure of the two sets of alleles has features consistent with this hypothesis, such as a block of silent changes in strong linkage disequilibrium with the Pro/Leu polymorphism. Furthermore, while Miyashita's four and six-cutter restriction map variation study of the 13 kb *G6pd* region found it to be one of the least variable regions yet studied,[26] our estimate of polymorphism for the silent sites was nearly fivefold higher.

The most striking aspect of the analysis was the level of interspecific divergence. We were expecting, using the levels observed for ADH, about 2-4 amino acid differences between *D. melanogaster* and *D. simulans*. However, there are 19 amino acid differences between the species, and overall, 21 replacement events can be inferred from the nucleotide sequence. We carried out two tests to determine the significance of this observation. Both tests use the contrast of intraspecific polymorphism against interspecific divergence for two classes of mutation (in this case, silent versus replacement changes). The first test, simply a G-test, as suggested by McDonald and Kreitman is highly significant ($G = 18.96$, $P << 0.001$), the second test, a variant of the conservative HKA test, is also significant ($X^2 = 7.9$, $P < 0.02$). This implies that there is a tenfold excess of replacement changes over that expected if the *G6pd* gene were evolving in a strictly neutral fashion.

It should be noted that the pattern of divergence between *D. melanogaster*, *D. simulans,* and rat and human G6PD was not particularly illuminating—it would never have predicted this dramatic rate. The rat and human enzymes, perhaps separated by 80 to 100 million years, show only 25 amino acid differences.[27] Furthermore, most of the differences between *D. melanogaster* and *D. simulans* occur clustered in about 20% of the sequence, and from simple inspection, there appears to be high congruence between sites of rat-human divergence and *melanogaster-simulans* divergence. Thus, most of the enzyme is strongly conserved, while the rest is turning over quickly in apparent adaptive response.

To examine this high rate of replacement change on a shorter scale, we are currently sequencing *D. mauritana* and *D. sechellia,* the close relatives of *D. simulans*. The results also show a number of replacement differences.

What Drives Adaptive Change at G6PD?

There are two potential scenarios for what causes adaptive amino acid substitution at G6PD. One is the simple Darwinian model. That is amino acid fixations are the selective response to varying (or maintain) pentose shunt function in changing environments. The second less conventional explanation might propose that many fixations reflect adaptive episodes that are in response to earlier amino acid fixations that confer reduced G6PD function (slightly deleterious mutation in the sense of Ohta[28]). Under this model mutations of reduced enzyme function fix as nearly neutral changes when population sizes are small, but become deleterious as population size expands. Since enzyme activity is now suboptimal, this may be functionally compensated for by (adaptive) amino acid changes that confer improved function. The scenario draws heavily on the arguments about limits to the evolution of enzyme function as discussed by Hartl, Dykhuizen and Dean.[29]

A preliminary examination of the polarity of fixed mutations at both silent and replacements in the separate *melanogaster* and *simulans* lineages (using *yakuba* as an outgroup) places twice as many replacement changes in the *simulans* lineage, while the converse is observed for synonymous changes. Since the prevailing view (based on polymorphism levels) seems to indicate a significantly larger historical population size for *simulans,* this would argue that most adaptive changes (i.e. replacements) are occurring in the larger population size lineage, while most slightly deleterious change (most of the silent changes are to disfavored codons) accumulate in the historically smaller *melanogaster* lineage. Therefore, the partitioning of G6PD changes between lineages supports Ohta's general model of nearly neutral mutation, but with the caveat (not embraced by Ohta) that much nearly neutral amino acid substitution must be adaptive.

Is G6PD Typical?

So far, "long term" patterns of fixation of amino acid replacements at PGI, 6PGD, GAPDH, and GPDH reflect molecules that are either very constrained, or appear to be changing in a neutral fashion when a McDonald-Kreitman configuration is tested. It is possible that the rather rapid rate of amino acid change at G6PD reflects its particular catalytic idiosyncrasies, or its position in the pathway architecture. Because its forward product is very unstable, G6PD is an effectively irreversible step, unlike most other enzymes in the pathway, and regulation of the pentose shunt is generally presumed to fall on G6PD.[30] Other than this distinction there are no other unique features. It should be pointed out that all the above enzymes, including G6PD, possess very high codon bias relative to other Drosophila genes (many of which are not enzymes). This strong bias suggests that these enzymes have evolved to "maximize" activity, and that small drops in activity, as would be associated with rare codon substitution, are detected in the face of drift by natural selection.

The issue of enzyme pathway architecture and step-specific propensities for adaptive change and polymorphism is an old theme, dating back to the idea of classifying enzymes into so-called regulatory and non regulatory groups[31] and examining the levels of allozyme polymorphism associated with each. This original idea, albeit intriguing, suffered from an inability to predict whether levels of polymorphism should be higher or lower at regulatory steps. This problem or prediction will not go away, however, the dimensionality added by being able to characterize and contrast intraspecific and interspecific variation at various classes or mutation (silent versus replacement), will allow individual tests for each regulatory or non-regulatory step in the pathway, with a subsequent overall test of generality. It would appear that this is an idea that should be revisited.

The Immediate Future—TPI, ALD, PGM, and HK

We are also beginning to study intra- and interspecific variation for triosephosphate isomerase (TPI). This enzyme is important because it acts to couple the branches between triglyceride metabolism and glycolysis. It, as G6PD, would be a responsive selective target for partitioning the rate of flux between alternate pathways. The two allele electrophoretic polymorphism in *D. melanogaster* is not well described, but our initial estimates indicate that it is absent, or rare, in African population and varies from 5–30% in temperate populations. The fast allele increases with latitude,[32] and therefore, covaries with the metabolically adjacent polymorphism at GPDH. No information exists on *in vivo* or *in vitro* differences between variant enzymes. The gene was recently cloned in *D. melanogaster*,[33] so it has been possible to begin PCR-facilitated sequencing. Our survey of twelve lines for a 1500 bp region spanning the *Tpi* gene is still incomplete, nevertheless we know that the polymorphism is due to a substitution of a glutamine for a lysine at amino acid residue 174. Whether the worldwide polymorphism is due to this change is unknown because at this time we have only sequenced the one fast electromorph. In addition, TPI has amino acid sequenced in a number of diverse taxa, and has long been an important model of enzyme catalysis.[34] Because it is one of the first proteins to have its three-dimensional structure determined,[35] we evaluate amino acid changes against a knowledge of functional importance. In fact, the polymorphism we have seen changes a lysine that is invariant among all taxa that have been sequenced, and involves a putative Lys-Glu salt bridge in the "hinged lid" over the active site. We also have plans to sequence and carry our a similar analysis on aldolase (ALD), whose sequence in *Drosophila melanogaster* was recently published.[36]

Acknowledgment

I acknowledge support from grants GM4524703 from the N.I.G.M.S and BSR9244885 from the N.S.F. Marty Kreitman, Chip Aquadro, Dave Begun and John McDonald kindly summarized and gave permission to discuss their unpublished results.

References

1. J. G. Oakeshott, J. B. Gibson, P. R. Anderson, W. R. Knibb, D. G. Anderson, and G. K. Chanbers. (1982). *Evolution* 36, 86–96.
2. J. G. Oakeshott, G. K. Chambers, J. B. Gibson, W. F. Eanes, and D. A. Willcocks. (1983). *Heredity* 50, 67–72.
3. R. S. Singh, and L. R. Rhomberg. (1987). *Genetics* 117, 255–271.
4. J. R. David, and P. CAPY. (1988). *Trends in Genetics* 4, 106–111.

5. D. Lachaise, M. Cariou, J. R. David, F. Lemeunier, L. Tsacas, and M. Ashburner, M. (1988). *Evolutionary Biology* 22, 163–225.
6. J. R. David, and C. Bocquet. (1975). *Nature* 257, 588–590.
7. W. J. Ewens. (1972). *Theor. Pop. Biol.* 3, 87–112.
8. W. J. Ewens. (1977). *Adv. Hum. Genet.* 8, 67–134.
9. J. H. Gillespie. (1991). *The Causes of Molecular Evolution*, Oxford University Press.
10. J. F. C. Kingman. (1982). *Stoch. Proc. Appl.* 13, 235–248.
11. N. L. Kaplan, T. Darden, and R. R. Hudson. (1988). *Genetics* 120, 819–829.
12. M. Kreitman. (1983). *Nature* 304, 412–417.
13. R. R. Hudson, M. Kreitman, and M. Aguade. (1987). *Genetics* 116, 153–159.
14. J. McDonald, and M. Kretiman. (1991). *Nature* 351, 652–654.
15. P. T. Barnes, and C. C. Laurie-Ahlberg. (1986). *Genetics* 112, 267–294.
16. G. C. Bewley, and S. G. Miller. (1979). In *Isozymes: Current Topics in Biological and Medical Research, (Vol. 3)* edited by (M. C. Rattazzi, J. G. Scanadalios, and G. S. Whitt, eds,) Alan Liss.
17. T. S. Takano, S. Kusakabe, and T. Mukai. (1993). In *Mechanisms of Molecular Evolution* (N. Takahata, and A. G. Clark, eds.). Sinauer, 179–190.
18. G. C. Bewley, J. L. Cook, S. Kusakabe, T. Mukai, D. L. Rigby, and G. K. Chambers. (1989). *Nucleic Acids Res.* 17, 8553–8567.
19. J. A. Coyne, W. F. Eanes, J. A. M. Ramshaw, and R. K. Koehn. (1979). *Syst. Zool.* 28, 164–175.
20. W. Geer, D. L. Lindel, and D. M. Lindel. (1979). Biochem. Genet. 17, 881–896.
21. W. F. Eanes, L. Katona, and M. Longtine. (1990). *Genetics* 125, 845–853.
22. J. Labate, and W. F. Eanes. (1992). *Genetics* 132, 783–787.
23. E. Beutler. (1991). *N. Engl. J. Med.* 324, 169–174.
24. W. F. Eanes, M. Kirchner, and J. Yoon. (1993). *Proc. Natl. Acad. Sci USA* 90, 7475–7479.
25. W. F. Eanes, and J. Hey. (1986). *Genetics* 113, 679–693.
26. N. T. Miyashita. (1990). *Genetics* 125, 407–419.
27. B. Persson, H. Jornvall, I. Wood, and J. Jeffrey. (1991). *Eur. J. Biochem.* **198,** 485–491.
28. T. Ohta. (1992). *Ann. Rev. Ecol. Syst.* **23**, 263–286.
29. D. L. Hartl, D. E. Dykhuizen, and A. M. Dean. (1985). *Genetics* **111**, 655–674.
30. T. Wood. (1985). *The Pentose Phosphate Pathway*, Academic Press.
31. G. B. Johnson. (1974). Enzyme polymorphism and metabolism. *Science* **184**, 28–31.
32. J. G. Oakeshott, S. W. McKechnie, and G. K. Chambers. (1984). *Genetica* **63**, 21–29.

33. R. L. Shaw-Lee, J. L. Lissemore, and D. T. Sullivan. (1991). *Mol. Gen. Genet.* 230, 225–229.

34. K. Ü. Yüksel, and R. W. Gracy. (1991). In *A study of enzymes II: Mechanism of enzyme action* (S. A. Kuby, ed.) pp. 457–484 CRC Press.

35. D. W. Banner, A. C. Bloomer, G. A. Petsko, D. C. Phillips, C. I. Pogson, I. A. Wilson, P. H. Corran, A. J. Furth, J. D. Milman, R. E. Offord, J. D. Priddle, and S. G. Waley. (1975). *Nature* 255, 609–614.

36. R. Shaw-Lee, J. L. Lissemore, D. T. Sullivan, and D. R. Tolan. (1992). Jour. Biol. Chem. 267, 3959–3967.

3

Molecular Population Genetics in *Drosophila Pseudoobscura*: Three Future Directions

Stephen W. Schaeffer

Abstract. *Nucleotide sequence data provides a powerful tool for population geneticists to discriminate between competing evolutionary hypotheses. The* Adh *region of* Drosophila pseudoobscura *has been a model system for the study of population genetic processes including mutation, random genetic drift, migration, recombination, and natural selection. Here, I summarize the relative importance of the major evolutionary forces acting on this relatively simple gene and suggest three future population studies that are motivated by the* Adh *data analysis.*

Introduction

One of the historical goals of population genetics was to estimate genetic variation in natural populations and to infer the evolutionary forces that have determined the pattern and organization of genomic diversity.[1] Previous attempts to make generalizations about the evolutionary process have been limited because of our inability to distinguish individuals that were identical-by-descent versus identical-by-kind. As a result, studies of genetic variation were consistent with multiple evolutionary hypotheses.[2,3] Two major advances in experimental and theoretical population genetics have provided powerful tools that allow alternative evolutionary scenarios to be discriminated. Population surveys of nucleotide sequences may now be collected as a result of the polymerase chain reaction and of rapid nucleotide sequencing methods.[4–6] Collections of nucleotide sequences may discriminate between hypotheses because the redundancy of the genetic code superimposes two histories on a nucleotide sequence, a selective history written in the amino acid sequences of proteins and a neutral history described by synonymous and noncoding sites.[7] In addition, low levels of recombination between nucleotides[8] correlate the inheritance patterns of linked selected and nonselected nucleotides. Thus, when balancing or directional selection acts on amino acid diversity, the number of variants in linked neutral sites may be increased or decreased.[2,3,9,10]

The statistical methods and theoretical models necessary to analyze and understand nucleotide sequence data have kept pace with or surpassed experimental studies of DNA in populations. Kimura's[11] neutral theory of molecular evolution is the null hypothesis that explains how nucleotide sequences will evolve when mutation, random genetic drift, migration and recombination are the major forces that act in populations. Critical population parameters and test statistics have been designed that are capable of rejecting expectations of the neutral theory.[12,13] Coalescent methods allow neutral gene genealogies to be simulated under various population genetic models without an extensive need for computer time and space.[14–21] Simulation results allow strong inferences to made about the past evolutionary history of populations. Thus, experimental and theoretical tools exist to determine how important natural selection has been in modulating genetic diversity in the genomes of several model organisms.

The alcohol dehydrogenase (*Adh*) region of *D. pseudoobscura* has been an experimental model of neutral sequence evolution within populations. The *Adh* study was originally undertaken to determine the evolutionary forces that influence the pattern and organization of nucleotide diversity at a phenotypically simple locus. I wanted to understand how a relatively simple locus evolved at the nucleotide level before I tried to examine sequences of genes whose evolutionary history was complicated by factors such as inversion polymorphisms, chromosomal context, or high degrees of amino acid variability. Population genetic parameters such as $4N\mu$, $4Nc$, and Nm have been estimated to determine if selection had acted on any of the nucleotide diversity, if the migration rate is large enough to homogenize gene frequencies among *D. pseudoobscura* populations, and if the recombination rate was equivalent to the mutation rate (N is the effective population size, μ is the mutation rate, c is the recombination rate and m is the migration rate). My laboratory has been able to answer these questions and address other issues that arose upon closer inspection of the data.

My laboratory has generated the complete nucleotide sequences of the 3.5 kilobase *Adh* region for 109 strains of *D. pseudoobscura* and its close relatives to infer the relative importance of mutation, random genetic drift, natural selection, migration, and recombination in the evolution of this region.[22–25] The large number of *D. pseudoobscura* strains were necessary to study higher order phenomena such as migration among populations and recombination among segregating sites. Ninety-nine strains were collected from 12 populations in the North American range of *D. pseudoobscura* that extends from British Columbia, Canada to Guatemala and from the Rocky Mountains to the Pacific Coast. $4N\mu$, $4Nc$, and Nm were estimated within nine of the 12 populations that had sample sizes greater than three strains (Fig. 1). Eight strains were collected from a disjunct population near Bogota, Colombia. The laboratory also determined single strains

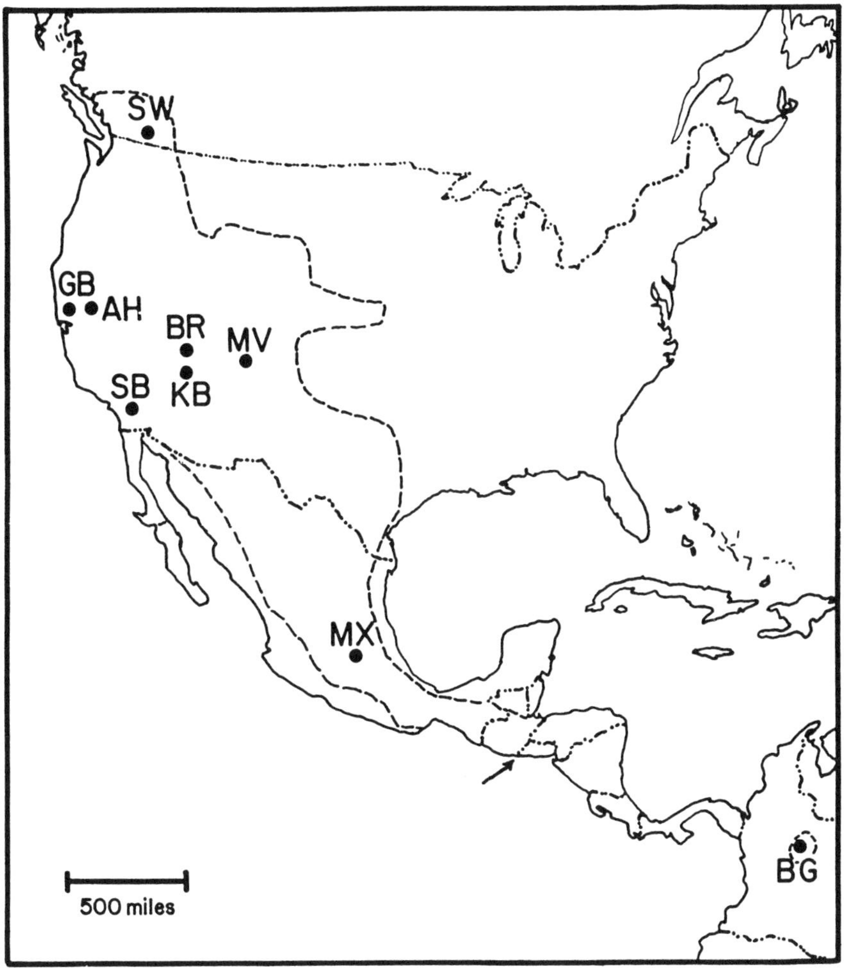

Figure 1. Geographic range and collection localities of *Drosophila pseudoobscura* with sample sizes greater than three. The geographic range is designated by (- - - -). The arrow shows the southern boundary of the range which is coincident with the southern border of Guatemala. The names of populations are: SW, Stemwinder Provincial Park, British Columbia, Canada; AH, Apple Hill, California; GB, Gundlach-Bundschu Winery, Sonoma Valley, California; BR, Bryce Canyon National Park, Utah; MV, Mesa Verde National Park, Colorado; KB, Kaibab National Forest, Arizona; SB, San Bernadino Mountains, California; MX, Mexico; and BG, Bogota, Colombia.

of the two sibling species, *D. persimilis* and *D. miranda*, to test segregating sites for departures from the neutral mutation hypothesis.

Selection in the Adh region

The *Adh* region is composed of two protein coding genes, *Adh* and *Adh-Dup*,[26] that differ in their levels of amino acid polymorphism (Fig. 2). Populations of *D. pseudoobscura* encode one major form of the ADH enzyme[27,28] and nine forms of the ADH-DUP enzyme (S. W. Schaeffer, Department of Biology, The Pennsylvania State University, unpublished data). What explains the differences in amino acid diversity in the two proteins? ADH could be monomorphic either because of a low neutral mutation at amino acid positions or because of a recent directional selection event that fixed a new beneficial allele. ADH-DUP could be polymorphic either because of a high neutral mutation rate at amino acid positions or because of some type of balancing selection maintaining alternative forms of the enzyme. I have discriminated between the neutral and selective hypotheses by an examination of linked neutral variation.[12,13] Levels of variation in linked neutral sites such as synonymous or intron sites may be decreased or increased by directional or balancing selection acting on amino acid variants.[20,29–32]

Two statistical tests of the neutral mutation hypothesis have been used to determine if variation in synonymous and intron sites departs from an equilibrium

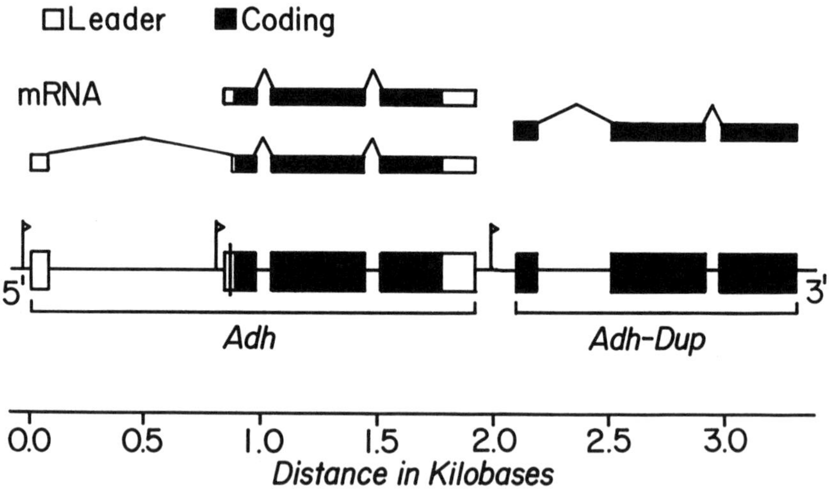

Figure 2. Fine structure of the *Adh* region of *D. pseudoobscura*. The region is subdivided into 17 sequence domains from 5′ to 3′: 5′ flanking, *Adh* adult leader, *Adh* adult intron, *Adh* larval leader, *Adh* exon 1, *Adh* intron 1, *Adh* exon 2, *Adh* intron 2, *Adh* exon 3, *Adh* 3′ leader, intergenic, *Adh-Dup* exon 1, *Adh-Dup* intron 1, *Adh-Dup* exon 2, *Adh-Dup* intron 2, *Adh-Dup* exon 3, and 3′ flanking.

neutral model.[12,13] Synonymous and intron sites are assumed to be selectively neutral, but the validity of this assumption may be violated as will be shown below. Neutral polymorphisms in *Adh* and *Adh-Dup* fail to reject an equilibrium neutral model as the cause of the observed amino acid variation.[22,24] The data provide no evidence for positive Darwinian selection acting on either of the two genes in the *Adh* region. The ADH enzyme is monomorphic because of a low neutral mutation rate at amino acid positions while ADH-DUP protein is highly polymorphic because of a higher neutral mutation rate at amino acid positions.

Positive Darwinian selection may not have acted on the amino acid variation in Adh or Adh-Dup, but other types of selection have left footprints in the sequences. Recent analyses of nonrandom associations among polymorphic sites in the *Adh* region and of the frequency spectra of *Adh* and *Adh-Dup* synonymous codons suggest that selection may discriminate between subtle phenotypic differences at the levels of pre-mRNA secondary structure and codon usage. Fisher's exact test of independence was used to test all pairs of variable nucleotide sites within the *Adh* region for nonrandom associations. No significant pattern of associations was discovered with the exception of two clusters of polymorphic sites that were nonrandomly associated in the adult intron and intron 2 of *Adh*.[23] A cluster of sites in linkage disequilibrium have also been observed in the *white* locus of *D. melanogaster*.[33,34] Nonrandom associations among genetic markers along a chromosome may be caused by low recombination rates, by population subdivision, by positive Darwinian selection with genetic hitchhiking, or strong epistatic selection. I have estimated the neutral recombination and mutation parameters in the Adh region and found that 7 to 17 recombination events occur for each mutation event.[23] Estimates of recombination rates do vary across the *Drosophila* genome,[35] however, there is no evidence that recombination rates can vary within a four kilobase region. Therefore, low recombination rates are unlikely to explain the clusters of linkage disequilibrium. Population subdivision, as will be shown below, may be ruled out because gene flow is sufficient to homogenize gene frequencies among all geographic populations of *D. pseudoobscura*.[22] As discussed above, I found no evidence for the action of positive Darwinian selection in the recent past of the Adh or the Adh-Dup genes.[22,24] Thus, strong epistatic selection seems to be the best explanation for the two clusters of linkage disequilibrium in the *Adh* introns. Why would selection favor particular combinations of segregating sites within introns? Stephan and Kirby[36] suggest that the clustered sites within the adult intron and intron 2 form stable stem-loop structures that are polymorphic within *D. pseudoobscura* populations. Recombination events that occur between alternative stem loops would not be favored and removed from the population. The linkage disequilibrium within the two clusters of segregating intron sites is maintained as a result of compensatory changes that retain stem-loop structures in the pre-mRNA of *Adh*.

Synonymous codons within genes are assumed to be selectively neutral because they do not change encoded amino acids in protein sequences. Nevertheless, the usage of synonymous codons in many species is nonrandom,[37–40] which suggests that selection may account for the observed biases. Hiroshi Akashi (Department of Ecology and Evolution, University of Chicago) and I wanted to examine the frequency spectra of new mutations that occur in preferred and in unpreferred codons to determine the fitness effects of theses synonymous mutations. A frequency spectrum is a distribution of gene frequencies for all polymorphic sites within in a gene region. The frequency spectrum will be U-shaped with most gene frequencies being zero or one if synonymous codons have equivalent fitness effects, the neutral expectation.[11] Frequency spectra that depart from the neutral case can have many shapes but two extreme cases result from purifying or balancing selection. Purifying selection acting on nucleotide variation will cause the frequency spectrum to have an excess of rare variants because selection prevents variants from reaching high frequencies. Balancing selection, on the other hand, will cause the frequency spectrum to have an excess of intermediate frequency variants. A frequency spectrum can be tested for departures from a neutral model with the Tajima test.[12]

The synonymous mutations were partitioned into two classes, PREFERRED where a preferred codon mutates to an unpreferred codon, or UNPREFERRED where an unpreferred codon changes to a preferred codon. The polarity of the mutations was determined from *D. miranda Adh* and *Adh-Dup* genes. The ancestral state of each codon, PREFERRED or UNPREFERRED, was determined by comparing the *D. miranda* codon with the codon bias table of 510 *Drosophila* genes.[41] The frequency spectra of *Adh* and *Adh-Dup* PREFERRED codons were significantly different from the neutral expectation and showed an excess of rare variants. On the other hand, the frequency spectra of *Adh* and *Adh-Dup* UNPREFERRED codons were consistent with a neutral model.[42] These results suggest that new mutations in preferred codons are slightly deleterious because purifying selection prevents these nucleotide changes from drifting to intermediate frequencies. Small effective population sizes lead to genes with low codon bias, which has been observed in regions of the *Drosophila* genome with reduced recombination.[43] Thus, the large effective population size of *D. pseudoobscura* probably leads to purifying selection capable of discriminating between UNPREFERRED and PREFERRED synonymous codons. All synonymous codons in *Drosophila* genes in regions of high recombination should use PREFERRED codons given the strong purifying selection against UNPREFERRED codons. This is not observed. Further data analyses are necessary to determine why some synonymous positions are selected and other silent codons are unselected.

Migration in D. pseudoobscura

A neutral genetic marker provides an excellent tool to estimate levels of migration between populations. The extent that flies can move between populations may be assessed with statistical tests that determine if genetic distance between individuals within populations is less than that observed between populations. Hudson *et al.*[44] suggest a random permutation method that tests the null hypothesis of no genetic difference between geographic populations. The nine populations of *D. pseudoobscura* were tested for significant population subdivision with the random permutation test of Hudson *et al.*[44] North American populations were found to be genetically uniform, while individuals in Bogota were shown to be genetically different from flies in North America.[22,25]

Nei's[45] measure of interpopulation nucleotide diversity was used to estimate the neutral migration parameter, *Nm*, among all populations in the North American range of *D. pseudoobscura.*[18,46] The estimate of *Nm* was 2.38 migrants between populations per generation[22] which is sufficient to homogenize gene frequencies between all populations of *D. pseudoobscura.*[47] In fact, evidence for extensive gene flow is dramatically demonstrated by the near identity of two sequences that were collected from British Columbia, Canada and Tulincingo, Mexico. The uniformity of allozymes frequencies is also consistent with extensive gene flow among populations.[1] Thus, any examples of genetic differentiation observed in *D. pseudoobscura* given such high levels of gene flow should be considered as strong evidence for the past action of natural selection.[22,48–50]

Molecular Evolution of Inversions Polymorphisms

The paracentric inversions of the third chromosome provide the most striking example of genetic differentiation in *D. pseudoobscura*. Inversion frequencies and the frequencies of allozymes carried by the inversions vary geographically and seasonally.[1,51–55] Thus, the selective forces imposed by different habitats and variable seasons overcome the extensive migration that acts to homogenize the frequencies of the inversions among populations. For example, Arizona populations are nearly fixed for the Arrowhead inversion while populations collected from Washington, Oregon, or California have at least four inversions at moderate frequency.[54,55] Immigration of Pacific coast flies should maintain inversion diversity in Arizona populations, however, selection seems to remove individuals that carry all inversion types except for the Arrowhead chromosome.

Pacific coast populations of *D. pseudoobscura* are highly polymorphic for third chromosomal inversions, however, the frequencies of the inversions vary widely among populations (Fig. 3). The Treeline (TL) inversion tends to be in high frequency in northern latitudes and in low frequency in southern populations, while the reverse is true of the Chiricahua (CH) chromosome. These data suggest that selection in heterogeneous environments is acting in coastal populations as it probably does in Arizona. What prevents inversions from becoming fixed in any of the populations in California, Oregon, or Washington? Dobzhansky showed that inversion frequencies in populations in the San Jacinto mountains cycle with the seasons.[53] This may suggest that the climate of the Arizona desert

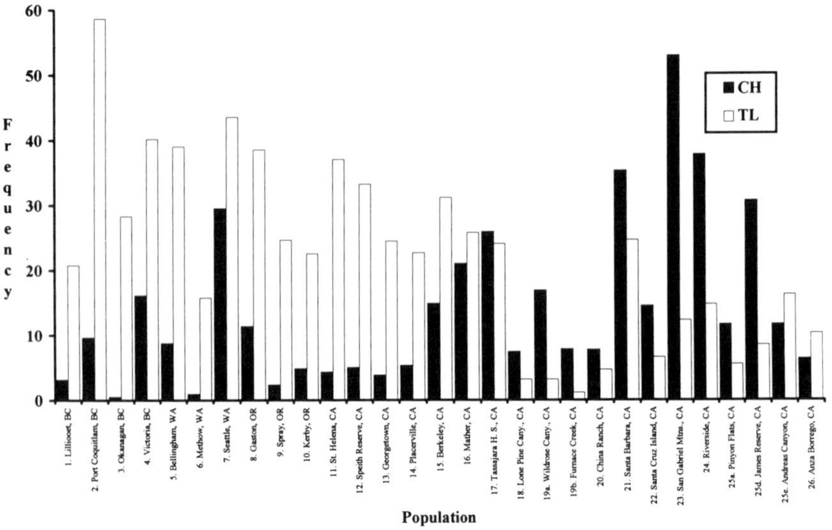

Figure 3. Frequencies of paracentric inversions in Pacific coast populations of *Drosophila pseudoobscura* collected during the 1980s (data taken from Anderson *et al.*[55]).

is relatively constant through time, while the environments of Pacific coast populations undergo regular climatic changes with each season. Powell and Anderson have reviewed field and laboratory experiments used to determine the basis of fitness differences of inversion homo- and heterozygotes.[56,57]

Studies of nucleotide sequences across the third chromosome of *D. pseudoobscura* have and will provide valuable insights about three issues in population genetics: (1) What is the nature of coadaption of genes within the inversions? (2) Do the genes within the inversions show evidence for selective sweeps? and (3) What impact does seasonal cycling of inversions have on the pattern and organization of nucleotide diversity? Each of these questions can be addressed by collecting sequence data for a sample of five scattered across the third chromosome collected from several *D. pseudoobscura* populations and data collected on the *Adh* region provides a natural control. Inversions have been considered a model of coadapted gene complexes because each inversion may have trapped a unique combination of genes that cannot be dissociated by recombination.[58] Studies of molecular variation may determine: (1) if the genetic information carried by the various inversions is similar or different; and (2) if the genetic information carried by a single inversion collected from geographic populations is similar or different.[59,60] Aquadro *et al.*[61] have shown that the origin of each inversion is monophyletic both within and between populations based on amylase nucleotide diversity, however, their data did not determine if nonrandom associations are maintained for genes across the chromosome both within and between populations. An examination of five gene sequences distributed across the inversions will determine if association among genes occurs within and between populations.

The effective amount of recombination for genes on the third chromosome should be inversely related to degree of inversion polymorphism observed within a population because inversion heterozygotes suppress recombination.[62] Regions of the *D. melanogaster* genome with low levels of recombination also tend to have low levels of genetic variation that may result from genetic hitchhiking associated with selective sweeps.[9,30,31,35,63] A study of nucleotide diversity in populations with low and high levels of inversion polymorphism provides a natural experiment to test the selective sweep model (Jeffrey R. Powell, Yale University, personal communication)(see other chapters of this volume for discussions of selective sweep models). Chromosomes sampled from populations with low levels of diversity would be expected to have higher levels of nucleotide variation than chromosomes collected from highly polymorphic populations.

Studies of nucleotide diversity of genes on the inversions may provide an experimental system to determine the effects of multiple selective sweeps, the so called traffic models. One of the puzzling observations made by Begun and Aquadro is that the frequency spectra for genes in regions of low recombination are not consistent with catastrophic sweeps where all variation is lost after a directional selection event. It has been suggested that these data may reflect multiple adaptive fixations that have occurred one after another with some overlap of events. These complex models of selective sweeps have been called traffic models. The seasonal shifts in inversion frequency observed by Dobzhansky,[53] suggest that traffic models might be examined experimentally with studies of gene evolution in populations that show yearly fluctuations of inversion frequencies. The frequency spectra of such genes may provide valuable insights about how frequency spectra change as a result of seasonal changes of the inversions.

Statistical Properties of Linkage Disequilibrium Parameters for Assemblages of Sites

Nucleotide sequence data may be used to detect selection largely because the action of natural selection at one site can create nonrandom associations or linkage disequilibria among polymorphisms at linked sites. The expected value of linkage disequilibrium for a pair of selectively neutral loci in finite equilibrium populations is zero, although the variance of linkage disequilibrium can be quite large between populations.[64–67] Ideally, one would like to have a statistical test that could determine if there is an excess of linkage disequilibrium above the neutral expectation. Estimates of pairwise linkage disequilibria have been used in past studies of restriction-map variation.[68–72] Nonrandom associations have been detected in these studies, however, none of the studies determined if the fraction of significant statistical tests of independence was higher than neutral expectation. One problem with comparing all variable sites in a region is the multiple comparison problem, if enough tests of independence are done than some may be significant by chance. A second problem is that segregating sites in a region may have different frequencies such that the power to detect significant associations among pairs of nucleotides is not equivalent.

A method that may overcome the multiple comparison and the power problems is to develop linkage disequilibrium statistics for an assemblage of segregating sites and determine the neutral expectation for the statistic. Brown *et al.* have developed statistics that summarize multilocus associations based on the moments of k, the number of pairwise differences.[73,74] Schaeffer and Miller have suggested that these multilocus statistics may be sensitive enough to pick up increased structure or association among segregating sites as a result of recent selection events.[24] For example, the *Adh* region in *D. melanogaster* has more associations among segregating sites due to the balanced polymorphism than the *Adh* region in *D. pseudoobscura*, which has little structure introduced by selection. Further work on the Brown *et al.* statistics might consider what the expectations of these statistics are given an infinite sites model of neutral evolution.

Molecular Population Genetics of Developmentally Regulated Genes

Population geneticists during this century have tried to determine how much genetic variation exists in natural populations and what was the functional significance of this variation.[1,75] The first question has largely been answered for a wide variety of organisms at the protein level.[76] The question of functional significance of observed variation is largely unanswered for many genes in plants and animals, although a recent compilation of restriction-map studies from *Drosophila* suggests that directional selection may have been a pervasive force in the genome.[35] Nucleotide sequence data do provide methods to determine which genetic loci have been acted upon by natural selection.[2,32,77] The latest goal of population genetics is to determine what fraction of the genome is acted upon by natural selection.[3]

I have chosen to study developmentally regulated genes of *Drosophila* as part of the ongoing effort to understand the impact that selection has had in evolution. The suite of genes that I will work on is part of the regulatory hierarchy that is responsible for sex determination and differentiation of *Drosophila*.[78–81] These genes are coordinately expressed such that genes expressed early in the hierarchy influence that expression of genes later in development. Sex determination in Drosophila can be divided into three stages, initiation, maintenance, and differentiation. The decision to become male or female is initially signaled by proteins that count the number of X and autosomal chromosomes.[79,81] The signal from the counting elements activates a central switch gene, *Sex-lethal*, in females that maintains the initial signal.[82] Sex-lethal and the next two genes in the hierarchy, *transformer* and *transformer-2*, regulate the female-specific expression of the *doublesex* gene, which turns on many other genes responsible for morphological differences between males and females.[78]

This cascade is an ideal model for the study of developmental evolution because the genes of the hierarchy are known from beginning to end, the genes have been sequenced, the genes are relatively small, and the genes fall into several functional classes. Thus, this survey offers the possibility to determine the types of selection, positive or negative, that act on developmental genes.

This study may determine if genes expressed early in development are more constrained than genes expressed later in development. Powell *et al.* have used DNA-DNA hybridization to show that embryonic mRNAs are more conserved than adult mRNAs.[83] They suggested that the synonymous changes were responsible for the observed differences, however, they had no direct evidence for their assertion. Early genes might be expected to be more conserved than late genes because loss-of-function mutations in early genes would tend to cause lethality, but not so with late genes. A survey of the sex determination genes will test the hypothesis that synonymous sites change more rapidly in late genes as opposed to early genes.

Molecular population genetic studies of sex determination genes may also examine the selective constraints on several types of regulatory genes. The functional importance of developmental genes would suggest that these proteins should be highly conserved, however, a study of sequences of the *transformer* locus between several *Drosophila* species shows some developmental genes can evolve at a rapid rate.[84] Three genes that encode counting elements expressed early in the hierarchy are basic Helix-Loop-Helix (bHLH) proteins that form homo- and heterodimers that may bind promoter sequences in DNA.[85–88] *Sex-lethal*, *transformer*, and *transformer-2* are RNA-binding proteins that alternatively splice pre-mRNAs in the sex determination pathway.[78] *Doublesex* is a transcription factor that coordinates the expression of sex-specific proteins.[89] Thus, a population genetic analysis will determine if different types of regulatory proteins are able to change within populations and if there is any evidence for positive Darwinian selection acting on developmental genes.

Computational Problems that Limit Productivity

The last area that I would like to mention has nothing to do with the science of population genetics but may be critical for any study that is currently undertaken in the field. One of the future challenges for molecular population genetics is to increase the ease that theoretical approaches and statistical tests are applied to experimental data and vice versa. Theoretical population geneticists want to estimate new parameters from sequence data and experimental population geneticists want to determine what evolutionary scenarios can explain variation patterns. I think that a number of computational and data base problems need to be solved if we are to increase our productivity in population genetics.

The first computational problem is that no databases exist for nucleotide sequences of homologous genes collected from populations. A researcher who would like to use published sequences must use one of three methods to obtain the data for analysis. The first method is to reconstruct the sequences from a list of polymorphic sites in a published paper because the data has not been entered into GenBank/EMBL Data Libraries. This is a time consuming process and the reconstructed data set may include many errors that may be introduced in the process. In addition, data in the published papers may not be converted into nucleotide sequences because authors fail to include the positions of segregating sites relative to an original published sequence. The problem is made worse when the sequences are difficult to align due to insertions and deletions.

Most authors do deposit their nucleotide sequences into the GenBank / EMBL Data Libraries so that published data are publicly available. One problem with the maintenance of population data in the data libraries is that it adds a large amount of redundant information that slows down all searches. The second method to obtain a collection of sequences is to download sequences from data bases to a personal computer through electronic methods. This insures that each sequence is correct, provided that it is free from errors in the data bases. The collection of sequences must then be aligned. Alignments of sequences collected from populations are generally not published so that the reconstructed alignment may not be consistent with that published in the journal. The last method to acquire sequences from populations is to obtain the aligned sequence data from the author.

A second computational problem is that each population geneticist uses a different format for their aligned sequence data file. Sequences could be in an interleaved format where the alignment of all sequences is presented in groups of 60 to 100 bases up to the total of N nucleotides. Alternatively, each sequence in its aligned form is listed sequentially in the data file. Some laboratories use numerical values for the nucleotides, while others encode the bases as characters. The aligned data file could contain information about the fine structure of the gene region as in eukaryotic genes. Other information about the sequences could also be included such as number of sequences drawn from a single population and the sequences that comprise each population. What is the best way to represent complex sequence organization such as introns, exons, and noncoding sequence? I have used a list of sequence features at the beginning of my sequence file so that I can extract information separately for each domain. The wide variety of formats used in different laboratories are not a serious problem because small computer programs can be written to interconvert file structures.

A third problem that has the potential to waste valuable brain power is the redundant construction of computer programs. There are two major classes of programs that I have written, sequence manipulation and parameter estimation programs. The first type of program enables subsets of data to be extracted from the set of aligned sequences. For instance, I have a program that will extract noncontiguous sequence domains and create a new file of aligned sequences. The second class of program can calculate important population genetic parameters such as $4N\mu$, 4Nc, or Nm. How many laboratories have also written computer programs to estimate important population genetic parameters? A great deal of effort is used to write programs and debug them so that correct estimates of parameters are obtained. In addition, all population geneticists are biased in their choice of programming language and computer platform. Standardization of sequence data files will make them compatible with all new programs if the programs use a standard data entry format. Perhaps a data base could then be established for computer programs at an FTP site so that all population geneticists could access the source code. I am not sure that now is the time to establish a program data base because I have not determined which programs have the most useful functions.

I am not suggesting that someone should compile a comprehensive set of programs that are driven by a slick menu driven system such as those used for phylogenetic analysis, PAUP (David Swofford, Smithsonian Institute) or PHYLIP (Joseph Felsenstein, University of Washington). The disadvantages of maintaining a suite of programs are that they are very labor intensive for the person that wrote the programs because the investigator spends many hours upgrading the programs and the creativity of other researchers is limited by the types of analyses the primary investigator chose to program. I think that a common data base for computer source code will allow all population geneticists to take advantage of the work done in other laboratories and will allow the nucleotide sequence data to be explored thoroughly.

The last problem for the analysis of populations of sequences is that computers may not always have enough memory to do simulations of neutral expectations because population parameters may be too large. This problem may be solved with new developments in computer hardware or new strategies for handling the sequence data sets with innovations in computer software. In the last decade, experimental population geneticists have had to generate sequences, apply appropriate tests to the data, and construct computer programs to analyze the data. I think that coordinating our efforts in the computational aspects of data analysis will increase our future productivity.

Acknowledgments

The sequence data described in this paper was determined by my research technician Ellen L. Miller without whose efforts these analyses would not have been possible. This work was supported in part by a grant from the National Institutes of Health (GM42472).

References

1. R. C. Lewontin. (1974). *The Genetic Basis of Evolutionary Change*, Columbia University Press.
2. M. Kreitman. (1987). *Oxf. Surv. Evol. Biol.* 4, 38–60.
3. M. Kreitman. (1991). in *Evolution at the Molecular Level* (R. K. Selander, A. G. Clark, and T. S. Whittam) 204–221, Sinauer.
4. R. K. Saiki, D. H. Gelfand, S. Stoffel, S. J. Scharf, R. Higuchi, G. T. Horn, K. B. Mullis, and H. A. Ehrlich. (1988). *Science* 239, 487–491.
5. R. G. Higuchi, and H. Ochman. (1989). *Nucleic. Acids Research* 17, 5865.
6. F. Sanger, S. Nicken, and A. R. Coulson. (1977). *Proc. Natl. Acad. Sci. USA* 74, 5463–5467.
7. R. C. Lewontin. (1985). *Ann. Rev. Genet.* 19, 81–102.
8. A. Chovnick, W. Gelbart, and M. McCarron. (1977). *Cell* 11, 1–10.
9. C. F. Aquadro. (1992). *Trends in Genet.* 8, 355–362.
10. C. F. Aquadro. (1993). In *Molecular Approaches to Fundamental and Applied Entomology* (Oakeshott, J. and Whitten, M.J.) 222–266, Springer-Verlag.
11. M. Kimura. (1983). *The Neutral Theory of Molecular Evolution*, Cambridge University Press.
12. F. Tajima. (1989). *Genetics* 123, 585–595.
13. R. R. Hudson, M. Kreitman, and M. Aguade. (1987). *Genetics* 116, 153–159.
14. J. F. C. Kingman. (1982). *J. Appl. Prob.* 19A, 27–43.
15. J. F. C. Kingman. (1982). *Stochast. Proc. Appl.* 13, 235–248.
16. F. Tajima. (1983). *Genetics* 105, 437–460.
17. R. R. Hudson. (1990). *Oxf. Surv. Evol. Biol.* 7, 1–44.
18. M. Slatkin. (1991). *Genet. Res.* 58, 167–175.
19. S. Tavare. (1984). *Theor. Pop. Biol.* 26, 119–164.
20. R. R. Hudson, and N. L. Kaplan. (1988). *Genetics* 120, 831–840.
21. N. L. Kaplan, T. Darden, and R. R. Hudson. (1988). *Genetics* 120, 819–829.
22. S. W. Schaeffer, and E. L. Miller. (1992). *Genetics* 132, 471–480.
23. S. W. Schaeffer, and E. L. Miller. (1993). *Genetics* 135, 541–552.
24. S. W. Schaeffer, and E. L. Miller. (1992). *Genetics* 132, 163–178.
25. S. W. Schaeffer, and E. L. Miller. (1991). *Proc. Natl. Acad. Sci. USA* 88, 6097–6101.
26. S. W. Schaeffer, and C. F. Aquadro. (1987). *Genetics* 117, 61–73.
27. G. K. Chambers, J. F. McDonald, M. McElfresh, and F. J. Ayala. (1978). *Biochem. Genet.* 16, 757–767.
28. S. Prakash. (1977). *Evolution* 31, 14–23.
29. D. J. Begun, and C. F. Aquadro. (1991). *Genetics* 129, 1147–1158.

30. M. Aguade, N. Miyashita, and C. H. Langley. (1989). *Genetics* 122, 607–615.
31. A. J. Berry, J. W. Ajioka, and M. Kreitman. (1991). *Genetics* 129, 1111–1117.
32. M. Kreitman, and R. R. Hudson. (1991). *Genetics* 127, 565–582.
33. N. Miyashita, and C. H. Langley. (1988). *Genetics* 120, 199–212.
34. N. M. Miyashita, M. Aguade, and C. H. Langley. (1993). *Genet. Res.* 62, 101–109.
35. D.J. Begun, and C. F. Aquadro. (1992). *Nature* 356, 519–520.
36. W. Stephan, and D. A. Kirby. (1993). *Genetics* 135, 97–103.
37. T. Ikemura. (1985). *Mol. Biol. Evol.* 2, 13–34.
38. P. M. Sharp, and W.-H. Li. (1987). *Mol. Biol. Evol.* 4, 222–230.
39. P. M. Sharp, and W.-H. Li. (1989). *J. Mol. Evol.* 28, 398–402.
40. D. C. Shields, P. M. Sharp, D. G. Higgins, and F. Wright. (1988). *Mol. Biol. Evol.* 5, 704–716.
41. K. Wada, Y. Wada, F. Ishibashi, T. Gojobori, and T. Ikemura. (1992). *Nucleic Acids Res.* 20, 2111–2118.
42. S. W. Schaeffer, and H. Akashi. (1994). *Nature* in preparation.
43. R. M. Kliman, and J. Hey. (1993). *Mol. Biol. Evol.* 10, 1239–1258.
44. R. R. Hudson, D. D. Boos, and N. L. Kaplan. (1992). *Mol. Biol. Evol.* 9, 138–151.
45. M. Nei. (1982). In *Human Genetics. Part A: The Unfolding Genome* (B. Bonne-Tamir, T. Cohen, and R. M. Goodman) 167–181, Alan R. Liss, Inc.
46. S. Wright. (1951). *Annals of Eugenics* 15, 323–354.
47. S. Wright. (1931). *Genetics* 16, 97–159.
48. J. A. Endler. (1973). *Science* 179, 243–250.
49. G. M. Simmons, M. E. Kreitman, W. F. Quattlebaum, and N. Miyashita. (1989). *Evolution* 43, 393–409.
50. A. Berry, and M. Kreitman. (1993). *Genetics* 134, 869–893.
51. T. Dobzhansky, and M. L. Queal. (1938). *Genetics* 23, 239–251.
52. T. Dobzhansky. (1939). *Genetics* 24, 391–412.
53. T. Dobzhansky. (1943). *Genetics* 28, 162–186.
54. T. Dobzhansky. (1944). *Carnegie Inst. Washington Publ.* 554, 47–144.
55. W. W. Anderson, J. Arnold, D. G. Baldwin, A. T. Beckenbach, C. J. Brown, S. H. Bryant, J. A. Coyne, L. G. Harshman, W. B. Heed, D. E. Jeffrey, L. B. Klaczko, B. C. Moore, J. M. Porter, J. R. Powell, T. Prout, S. W. Schaeffer, J. C. Stephens, C. E. Taylor, M. E. Turner, G. O. Williams, and J. A. Moore. (1991). *Proc. Natl. Acad. Sci. USA* 88, 10367–10371.
56. W. W. Anderson. (1989). *Genome* 31, 239–245.
57. J. R. Powell. (1992). In *Drosophila Inversion Polymorphism* (C.B. Krimbas, and J. R. Powell) 73–126, CRC Press.

58. T. Dobzhansky. (1970). *Genetics of the Evolutionary Process*, Columbia University Press.

59. T. Dobzhansky. (1948). *Genetics* 33, 588–602.

60. T. Dobzhansky, and O.A. Pavlovsky. (1953). *Evolution* 7, 198–210.

61. C. F. Aquadro, A. L. Weaver, S. W. Schaeffer, and W. W. Anderson. (1991). *Proc. Natl. Acad. Sci. USA* 88, 305–309.

62. M. Ashburner. (1989). *Drosophila: A Laboratory Handbook*, Cold Spring Harbor Laboratory Press.

63. M. Aguade, N. Miyashita, and C. H. Langley. (1989). *Mol. Biol. Evol.* 6, 123–130.

64. W. G. Hill, and A. Robertson. (1968). *Theor. Appl. Genet.* 38, 226–231.

65. T. Ohta, and M. Kimura. (1969). *Genetics* 63, 229–238.

66. G. B. Golding. (1984). *Genetics* 108, 257–274.

67. R. R. Hudson. (1985). *Genetics* 109, 611–631.

68. C. F. Aquadro, R. M. Jennings, M. M. Bland, C. C. Laurie, and C. H. Langley. (1992). *Genetics* 132, 443–452.

69. C. F. Aquadro, S. F. Deese, M. M. Bland, C. H. Langley, and C. C. Laurie-Ahlberg. (1986). *Genetics* 114, 1165–1190.

70. C. H. Langley, A. E. Shrimpton, T. Yamazaki, N. Miyashita, Y. Matsuo, and C. F. Aquadro. (1988). *Genetics* 119, 619–629.

71. S. W. Schaeffer, C. F. Aquadro, and W. W. Anderson. (1987). *Mol. Biol. Evol.* 4, 254–265.

72. S. W. Schaeffer, C. F. Aquadro, and C. H. Langley. (1988). *Mol. Biol. Evol.* 5, 30–40.

73. A. H. D. Brown, M. W. Feldman, E. and Nevo. (1980). *Genetics* 96, 523–536.

74. A. H. D. Brown, and M. W. Feldman. (1981). *Proc. Natl. Acad. Sci. USA* 78, 5913–5916.

75. T. Dobzhansky. (1955). *Cold Spring Harbor Symp. Quant. Biol.* 20, 1–15.

76. E. Nevo. (1978). *Theor. Pop. Biol.* 13, 121–177.

77. A. L. Hughes, and A. Nei. (1988). *Nature* 335, 167–170.

78. B. S. Baker. (1989). *Nature* 340, 521–524.

79. T. W. Cline. (1985). In *The Origin and Evolution of Sex* (H. O. Halvorson, and A. Munroy) 301–327, Alan R. Liss, Inc.

80. M. Steinmann-Zwicky. (1988). *EMBO J.* 7, 3889–3898.

81. S. M. Parkhurst, and D. Ish-Horowicz. (1992). *Current Biology* 2, 629–631.

82. L. N. Keyes, T. W. Cline, and P. Schedl. (1992). *Cell* 68, 933–943.

83. J. R. Powell, A. Caccone, J. M. Gleason, and L. Nigro. (1993). *Genetics* 133, 291–298.

84. M. T. O'Neil, and J. M. Belote. (1992). *Genetics* 131, 113–128.

85. C. Murre, P. S. McCaw, and D. Baltimore. (1989). *Cell* 56, 777–783.

86. C. Murre, P. S. McCaw, H. Vaessin, M. Caudy, L. Y. Jan, Y. N. Jan, C. V. Cabrera, J. N. Buskin, S. D. Hauschka, A. B. Lassar, H. Weintraub, and D. Baltimore. (1989). *Cell* 58, 537–544.

87. M. Torres, and L. Sanchez. (1989). *EMBO J.* 8, 3079–3086.

88. S. M. Parkhurst, D. Bopp, and D. Ish-Horowicz. (1990). *Cell* 63, 1179–1191.

89. K. C. Burtis, K. T. Coschigano, B. S. Baker, and P. C. Wensink. (1991). *EMBO J.* 10, 2577–2582.

4

Selection, Recombination, and DNA Polymorphism in *Drosophila*

Charles F. Aquadro, David J. Begun and Eric C. Kindahl

Introduction

A goal of molecular population genetics is to provide an historical understanding of evolutionary processes occurring within and between closely related populations. While molecular techniques can, in principle, be applied to any species, the use of a model system such as *Drosophila melanogaster* has proved to be enormously fruitful. One recent finding demonstrates the utility of using a well characterized genetic system: Levels of (presumably) neutral DNA variation are positively correlated with recombination rates in *D. melanogaster*.[1] In this chapter we discuss recent results from our lab which extend this early result, discuss competing models to explain the pattern, and discuss empirical approaches to distinguish among these models.

Population genetic models

The neutral allele theory of molecular evolution posits that most new mutations are strongly deleterious and are kept in low frequency or rapidly eliminated by purifying selection. The variation we observe within and between species has no fitness consequences. These neutral variants arise at random, and drift to loss or fixation by a stochastic process.[2] Variation within a population is a function of population size and mutation rate. At equilibrium, heterozygosity per nucleotide is expected to equal $4N\mu$, where N is the effective (breeding) population size of the organism and μ is the mutation rate to functionally equivalent variants. The amount of sequence divergence between species is a simple function of the time since divergence and the mutation rate. Under this model, gene regions experiencing a high mutation rate and/or low functional constraint are expected to be highly variable within species and highly divergent between species. In general, the model predicts a positive correlation between variation within species and divergence between species.

Bringing adaptive substitutions into the picture has important population genetic consequences. Directional selection will drive a variant through a population to fixation while eliminating much of the neutral variation at linked sites. The size of genomic region affected by this "hitchhiking effect" depends on several factors, including the population size, the selection coefficient and rate of crossing-over between the selected site and neighboring neutral sites.[3–5] These models predict that the size of the genomic region over which variation is reduced increases as the recombination rate decreases. Importantly, the hitchhiking process is not expected to affect sequence divergence between species at linked, neutral sites,[6,7] resulting in an uncoupling of within and between species evolution (contrary to the prediction of neutral models).

Levels of DNA Variation Correlate with Recombination Rate in *D. melanogaster*

Genomic regions experiencing low crossing-over in *Drosophila* tend to have low variability compared to regions of normal crossing-over[8–10] but show normal levels of interspecific divergence, a pattern which was interpreted as evidence of hitchhiking.[11–14] Furthermore, we found a general relationship between crossing-over and DNA polymorphism, suggesting that hitchhiking effects were reducing variation across much of the *D. melanogaster* genome.[1] This observation was intriguing, but the data were somewhat limited. Specifically, we thought it was important to collect data from a single geographic region and from a large number of genes experiencing different recombination rates, from high to intermediate and down to severely reduced (e.g. centromeric regions of *Drosophila*). We hoped to better quantify the shape of the relationship, to determine if the level of DNA polymorphism varies linearly or curvilinearly with crossing-over, and to compare autosome and *X*-linked variability.

DNA heterozygosity was measured with a high-resolution four-cutter technique over 2–5 kb segments of 15 genetically well mapped and physically localized genes distributed across the third chromosome in a sample of 96 *D. melanogaster* lines collected in Beltsville, Maryland.[15] These genes are *Lsp1-γ, Hsp26, Sod, Est6, fz, tra, Pc, Antp, Gld, ry, Ubx, Rh3, E(spl), Tl and Mlc2*. Recombination rates vary dramatically across the third chromosome (Figure 1) which constitutes approximately 40% of the *D. melanogaster* genome.

In agreement with the results of Begun and Aquadro,[1] nucleotide variability was highly correlated with regional rates of recombination (Figure 2). These data provide strong support for the notion that recombination rate significantly affects DNA polymorphism across the *entire range of estimated recombination rates* and that for a North American population of *D. melanogaster*, variation

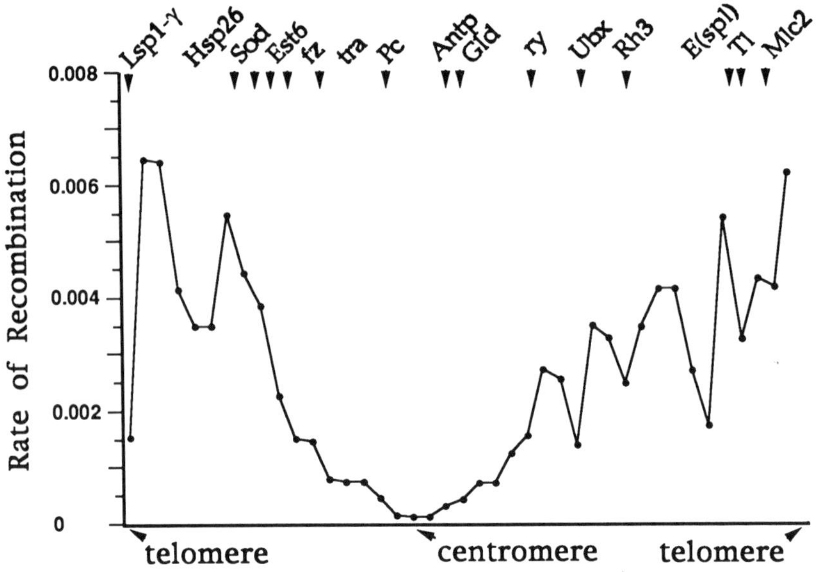

Figure 1. Plot of rates of recombination across chromosome III with the locations of 15 gene regions surveyed for variation indicated (modified from ref. 15). Rates of recombination were calculated as centimorgans (cM) per cytological band, averaged for intervals of 20–50 bands for markers flanking the gene region of interest (from ref. 39). Values plotted here represent rates of recombination estimated for natural populations and thus represent cM/band multiplied by 1/2 (this assumes a 1:1 sex ratio and reflects the fact that there is normally no recombination in males).

in crossing-over explains a significant amount of the heterogeneity in DNA sequence variation.

Why Does the Correlation Exist?

The reduction in polymorphism associated with reduced crossing-over has typically been ascribed to hitchhiking effects of advantageous mutations. Recently, however, a second possibility has been proposed. Charlesworth *et al.*[16] showed that removal of strongly deleterious mutation by selection can reduce levels of linked, neutral variation (subsequently referred to as "background selection") and that the effect is most pronounced for regions of reduced crossing-over. Thus, background selection and hitchhiking make qualitatively similar predictions of a positive correlation between polymorphism and crossing-over. Since deleterious mutations and advantageous mutations occur in natural populations, both background selection and hitchhiking occur. The challenge is to determine how (or whether) the models make different, testable predictions so that we can

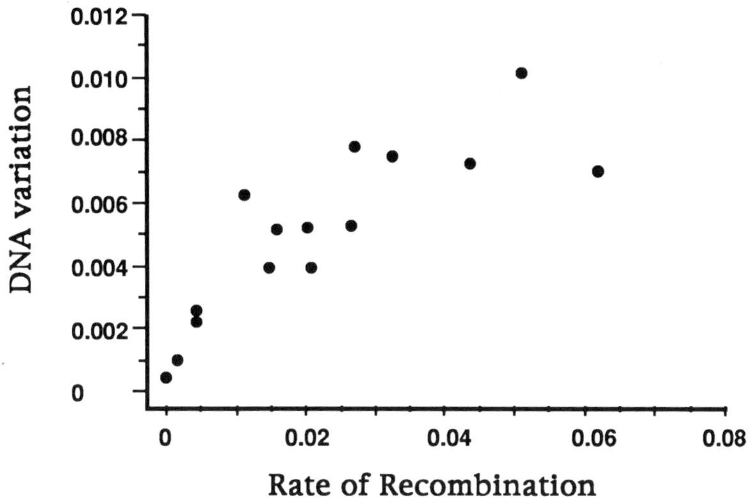

Figure 2. Scatterplot of nucleotide variation (θ ref. 40) versus rate of recombination (calculated as described in the Figure 1 legend) for 15 third chromosome gene regions (modified from Kindahl and Aquadro[15]).

determine the relative contributions of hitchhiking and background selection to the observed correlation between variation and recombination. One potentially important difference between the models is their predictions for levels of DNA variation in regions of equivalent recombination on the *X* chromosome versus autosomes (after taking into account the effects on variation and recombination due to differences in effective population size between *X*-linked and autosomal genes).

Since *X*-linked genes are hemizygous in males, recessive advantageous mutations are fixed more rapidly on the *X* chromosome than on the autosomes.[17] Because the magnitude of the hitchhiking effect depends, in part, on the fixation rate of selected mutations, hitchhiking models may predict that variation will be more severely reduced for *X*-linked genes compared to autosomal genes, all else being equal.

In contrast, background selection is a mutation-selection equilibrium model in which recessive deleterious mutations are eliminated more efficiently from the *X* chromosome than the autosome because of hemizygous males; the *X* chromosome has a larger proportion of mutation-free gametes compared to the autosomes. Because the effective population size of the mutation-free class of chromosomes determines the amount of variation,[16] *X*-linked genes are expected to have higher levels of variation (all else being equal) after correcting for different population sizes of sex-linked and autosomal genes.[1]

Kindahl and Aquadro[15] have made a preliminary test of these predictions using their four-cutter restriction site data from the third chromosome and additional data from seven *X*-linked genes assayed in other North American populations by us or by other investigators (summarized in ref.18). Levels of variation and recombination for *X*-linked genes have been adjusted to make them comparable to autosomal genes (see Figure 3 legend). Figure 3 shows that for genes experiencing similar recombination rates, sequence variation is generally lower for *X*-linked genes than for third chromosome genes, supporting the notion that an important cause of the correlation between recombination and variation is the hitchhiking effect of advantageous substitutions. There was also a hint of lower variation for *X*-linked genes in the data originally compiled in Begun and Aquadro.[1] A more quantitative analysis of this pattern is underway.

How Much Advantageous Selection?

Assume for the moment that we can ignore background selection as a major cause of the correlation. How much selection would be required to explain the

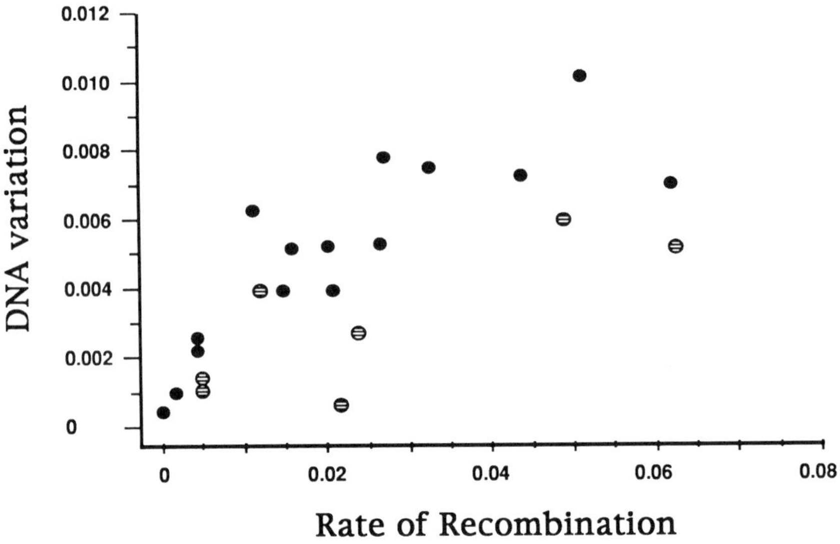

Figure 3. Scatterplot of nucleotide variation (θ) versus rate of recombination (adjusted coefficient of exchange) for 15 third-chromosome gene regions and seven X-linked gene regions, all from populations from the United States of America. Third-chromosome and X-linked genes are represented by closed and hatched circles, respectively. The third chromosome data are from Kindahl and Aquadro[15] and the X-chromosome data are from the literature (summarized in Begun and Aquadro[1]). Levels of variation for X-linked genes have been adjusted to make them comparable to autosomal genes by multiplying the former by 4/3. Rates of recombination, have been adjusted (see Figure 1 legend) by 1/2 for autosomes, and by 2/3 for X-chromosomes to compensate for the lack of recombination in males and assuming a 1:1 sex ratio.

observed relationships between polymorphism and crossing-over? The slope of variation vs. crossing-over may allow estimation of the number and intensity of selective events assuming that selection has occurred at a relatively constant rate over the history of the *D. melanogaster-D. simulans* lineage.[19,20] Applying the method of Wiehe and Stephan[20] to our third-chromosome data in Figure 2 yields an estimate of the intensity of selective sweeps ($2Ns\nu$) of 5.4×10^{-8} where N, s, and v are the effective population size, the selection coefficient, and the number of strongly selected substitutions per nucleotide site per generation, respectively.[15] This is identical to the estimate of $2Ns\nu$ obtained by Wiehe and Stephan[20] using the data summarized in Begun and Aquadro.[1] The asymptotic level of variation in regions of highest recombination suggests that $H_{neutral}$ (an estimate of $4N_e\mu$) is approximately 0.008. If $\mu = 1 \times 10,^{-9}$ then N is estimated to be $2 \times 10.^6$ Assuming an average selective effect (s) of 0.01 and the above estimate of N, then the estimated fixation rate of advantageous mutants per base pair (ν) is $1.4 \times 10.^{-12}$ This model leads to the (admittedly) crude estimate that 1 out of every 740 substitutions are driven to fixation by selection. We can use this model to get a sense for the number and frequency of advantageous substitutions.

D. melanogaster and *D. simulans* differ on average by 5% of their nucleotides.[21] The genome size for nonrepetitive sequences is 1.15×10^8 bp in both species. Thus, 5.75×10^6 fixations have occurred between these two species (due to drift and selection). The two species diverged approximately 2.5×10^6 years ago, so the total time in the *D. melanogaster-D. simulans* lineage is about 5×10^6 years. Therefore, a fixation has occurred (on average) every 1.15 years. If 1/740 were selectively driven, then we estimate about one selective sweep every 850 years (or perhaps 8500 generations). Therefore, about $(1/740)(1.15 \times 10^8\text{bp})(0.05)$ or 7770 selectively fixed differences separate *D. melanogaster* and *D. simulans*. If there are about 1×10^4 genes in the *D. melanogaster* genome, an average gene is about 300 amino acids long and has 1% amino acid divergence between *D. melanogaster* and *D. simulans*, then the total number of amino acids coded in the genome is 3×10^6 and the total number of amino acid differences between species is about $3 \times 10.^4$ If all selectively fixed differences between species are amino acid differences then the percentage of amino acid differences fixed by adaptive substitutions is about 7770 divided by 3×10^4 or 26%. Alternatively, assuming the Wiehe and Stephan model is approximately correct, if the average selection coefficient is smaller than 0.01 then we have underestimated the number of selectively driven amino acid replacements. On the other hand, if we are incorrect in assuming that most adaptive fixations are amino acid differences (e.g., if regulatory mutations are very important) then we may have overestimated the number of amino acid differences resulting from selection. If our estimates are realistic then a large fraction of protein evolution may be driven by natural selection.[22] Interestingly, two of the six genes analyzed with the McDonald-Kreitman test[23] in species of the *D. melanogaster* subgroup have shown departures from neutrality consistent with adaptive amino acid replacement differences between species.[23–27] This level of selection is in line with the expectation from the estimates using polymorphism and crossing-over data in *D. melanogaster*. Of course, these estimates are consistent with the notion that the majority of total nucleotide differences between species results from neutral mutations and genetic drift.[2] Obviously, we have relied on a great many assumptions. Refinements of both theory and data may allow more precise estimation.

Genetic Differentiation between *Drosophila* Populations

If selective sweeps are as common as the above calculations would indicate, geographically distant populations might be fixed for different sequence variants in regions of reduced recombination provided that migration is small relative to differential advantageous selection. Such a pattern has been reported by Stephan and Mitchell[10] for *D. ananassae* and by us[18] for *D. melanogaster* samples from the United States and Zimbabwe. In both species, regions of reduced crossing-over had low variation within populations and nearly fixed differences between populations. This pattern may be partly explained by independent hitchhiking events.[10,18]

Could background selection contribute to increased differentiation in regions of reduced crossing-over? The effect of background selection is to reduce effective population size in regions of low recombination. Since the variance of allele frequencies is greater in smaller populations, background selection might accelerate differentiation in regions of reduced crossing-over. Charlesworth *et al.*[16] estimate that N is reduced to 76% of the strictly neutral level in the distal portion of the *X* chromosome. $F_{ST} = 1/(3Nm + 1)$ for *X*-linked genes at equilibrium, where N is the effective population size and m is the migration rate. Begun and Aquadro[18] estimated F_{ST} to be about 0.30 and 0.60 for *X*-linked genes in regions of high recombination and low recombination, respectively. These values of F_{ST} lead to estimates of Nm that differ by 3.5-fold, a much greater difference than that predicted for background selection. This line of reasoning again supports the importance of selective sweeps in determining levels of intra- and interpopulation variation at the DNA sequence level. A pressing theoretical need is rigorous quantitative analysis of background selection and hitchhiking with population subdivision.

Inconsistencies Between Data and Models

Although the observed patterns regarding overall heterozygosity versus recombination are qualitatively consistent with an important contribution of hitchhiking, certain aspects of the data are troubling. For example, under the simplest "catastrophic" selective sweep model with no recombination, neutral variation is eliminated at the time of fixation of the linked, selected site. As such regions recover from monomorphism (or near monomorphism), the frequency distribution of variation is expected to be skewed towards rare variants and approach the neutral distribution relatively slowly.[7,29,30] However, restriction site data from *D. melanogaster* show no consistent skewness of the frequency spectrum in regions of severely reduced recombination.[15,18,31] A potential explanation for the absence of a skewed distribution is that rare variants present in an ancestral population were lost during the founding of recently established populations.[32] However, the frequency distribution in Zimbabwe in regions of low crossing-over do not support this hypothesis.[18] Regions of reduced crossing-over experiencing balancing selection, overlapping selective sweeps (favoring different variants in trans phase on chromosomes), or deleterious mutations linked to advantageous mutations could alter the distribution of variation in complicated ways, and represent fertile ground for theoretical and empirical investigation.

Is there any evidence for balancing selection in regions of reduced crossing-over? While DNA heterozygosity is reduced in these genomic regions, allozyme variation (heterozygosity and number of alleles per locus; refs. 33–35) in *D. melanogaster* shows no correlation with crossing-over.[15] Hitchhiking in regions of low crossing-over might remove much neutral polymorphism while strongly selected balanced polymorphisms persist (cf. ref. 36). In any case, we think direct surveys of replacement and silent variation in regions of reduced crossing-over will be worthwhile, but more sophisticated models will probably be required to interpret such data.

Conclusions

There are often two competing interests in population genetics, testing evolutionary models and estimating population parameters. The correlation between polymorphism and recombination is important in furthering both interests.

Though it is premature to make strong statements asserting the predominant role of hitchhiking as opposed to background selection, the available data are more favorable for the hitchhiking model; this includes both the correlation between variation and crossing-over as well as the increased differentiation between populations in regions of reduced crossing-over. Thus we conclude that advantageous selection is sufficiently common to have a major effect on neutral sites across the genome. Since hitchhiking reduces the effective population size of a genomic region, it has important effects on any evolutionary process dependent on population size. For example, slightly deleterious mutations are more likely to make a contribution to heterozygosity and divergence in regions of reduced crossing-over since selection is less effective in such regions.[37] This could have real consequences for interpreting patterns of silent and replacement variation in different species. An example from our lab is instructive. Aquadro *et al.*[38] interpreted the patterns of variation at the *rosy* locus in *D. melanogaster* and *D. simulans* as evidence for a model in which most protein polymorphism is slightly deleterious and the effective population size of *D. simulans* is several fold greater than that of *D. melanogaster*. Now we must at least consider whether the contrasting patterns of silent and replacement variation may in part, be a result of a different recombinational environment for the *rosy* locus region in the two species and thus different effective sizes for *rosy* rather than for the entire species. We are currently testing these possibilities.

Hitchhiking has important implications if one hopes to infer effective population size from DNA data. Specifically, one cannot reliably do so without prior knowledge of the relationship between crossing-over and variation in the species of interest or without invoking a specific selection model (e.g. ref. 20) which may or may not be correct. Pervasive hitchhiking could reduce genome variability in a wide range of taxa, providing a reasonable explanation for the limited upper range of protein heterozygosity observed across diverse species, as originally proposed by Maynard Smith and Haigh[3] and reiterated by Gillespie.[22] This also raises the interesting possibility that the recombinational environment of a gene is an important determinant of the rate and pattern of evolution between species.

Could hitchhiking effects of advantageous mutations contribute to the evolution of reproductive isolation? Slightly deleterious alleles could conceivably be fixed between populations in genomic regions where hitchhiking effects are particularly strong. In this way, a secondary effect of strong advantageous selection could be alternative fixation of linked alleles with phenotypic effects (as well as from pleiotropic effects of the advantageous alleles). If the major cause of isolation is the pleiotropic effect of advantageous alleles themselves, then genes causing isolation should be found distributed across the genome. On the other hand, if hitchhiking of alleles at linked sites is the major cause of isolation, then genes responsible for isolation should be found predominantly in regions of reduced crossing-over. These predictions may be experimentally testable.

Molecular population genetic analysis of multiple loci has been a useful approach for demonstrating the importance of the interaction between positive selection and crossing-over within and between closely related populations. Given the rapid pace of technological development for genome mapping, it is not inconceivable that the basic framework for exploring these patterns could be extended to a wide range of taxa. Further empirical and theoretical exploration of the relationship between crossing-over and DNA sequence variation will probably be an exciting part of population genetics for several years.

Acknowledgments

We are grateful to Brian Golding, Richard Hudson and Michael Nachman for commenting on this chapter. Research in our laboratory is supported by grants from the NIH and NSF.

References

1. D. J. Begun, and C. F. Aquadro. (1992). *Nature* 356, 519–520.
2. M. Kimura. (1983). *The Neutral Theory of Molecular Evolution,* Cambridge University Press.
3. J. Maynard Smith, and J. Haigh. (1974). *Genet. Res.* 23, 23–35.
4. N. L. Kaplan, R. R. Hudson, and C. H. Langley. (1989). *Genetics* 123, 887–899.
5. W. Stephan, T. H. E. Wiehe, and M. W. Lenz. (1992). *Theor. Pop. Biol.* 41, 237–254.
6. C. W. Birky Jr., and J. B. Walsh. (1988). *Proc. Natl. Acad. Sci U.S.A.* 85, 6414–6418.
7. R. R. Hudson. (1990). *Oxf. Surv. Evol. Biol.* 7, 1–44.
8. M. Aguadé, N. Miyashita, and C. H. Langley. (1989). *Genetics* 122, 607–615.
9. W. Stephan, and C. H. Langley. (1989). *Genetics* 121, 89–99.
10. W. Stephan, and S. J. Mitchell. (1992). *Genetics* 312, 1039–1045.
11. D. J. Begun, and C. F. Aquadro. (1991). *Genetics* 129, 1147–1158.
12. A. J. Berry, J. W. Ajioka, and M. Kreitman. (1991). *Genetics* 129, 1111–1117.
13. J. M. Martín-Campos, J. M. Comerón, N. Miyashita, and M. Aguadé. (1992).
14. C. H. Langley, J. MacDonald, N. Miyashita, and M. Aguadé, M. (1993). *Proc. Natl. Acad. Sci. USA* 90, 1800–1803.
15. E. C. Kindahl, and C. F. Aquadro. (unpublished).
16. B. Charlesworth, M. T. Morgan, and D. Charlesworth. (1993). *Genetics* 134, 1289–1303.
17. B. Charlesworth, J. A. Coyne, and N. H. Barton. (1987). *Amer. Nat.* 130, 113–146.
18. D. J. Begun, and C. F. Aquadro. (1993). *Nature* 356, 548–550.
19. C. F. Aquadro, and D. J. Begun. (1993). In *Molecular Paleo-population Biology* (N. Takahata, and A. G. Clark, eds) Japan Sci. Soc. Press/Sinauer.
20. T. H. E. Wiehe, and W. Stephan. (1993). *Molec. Biol. Evol.* 10, 842–854.
21. C. F. Aquadro. (1993). *Trends Genet.* 8, 355–362.
22. J. H. Gillespie. (1991). *The Causes of Molecular Evolution*, Oxford University Press.
23. J. H. McDonald, and M. Kreitman. (1991). *Nature* 351, 652–654.
24. W. F. Eanes, M. Kirchner, and J. Yoon. (1993). *Proc. Natl. Acad. Sci. USA* 90, 7475–7479.
25. R. M. Kliman, and J. Hey. (1993). *Genetics* 133, 375–387.
26. J. Hey, and R. M. Kliman. (1993). *Molec. Biol. Evol.* 10, 804–822.
27. F. J. Ayala, and D. L. Hartl. (1993). *Molec. Biol. Evol.* 10, 1030–1040.
28. T. Maruyama, and P. A. Fuerst. (1984). *Genetics* 108, 745–763.

29. C. H. Langley. (1990). In *Population Biology of Genes and Molecules* (N. Takahata, and J. F. Crow) Baifukan Co. Ltd., Tokyo.
30. F. Tajima. (1993). in *Molecular Paleo-population Biology* (N. Takahata, and A. G. Clark, eds) Japan Sci. Soc. Press/Sinauer.
31. M. Aguadé, M. Meyers, A. D. Long, C. H. Langley. (in press). *Proc. Natl. Acad. Sci. USA.*
32. J. R. David, and P. Capy. (1988). *Trends Genet.* 4, 106–111.
33. R. S. Singh, D. A. Hickey, and J. R. David. (1982). *Genetics* 101, 235–256.
34. M. Choudhary, and R. S. Singh. (1987). *Genetics* 117, 697–710.
35. R. S. Singh, and L. R. Rhomberg. (1987). *Genetics* 117, 255–271.
36. J. Sved. (1977). In *Proceedings of the International on Quantitative Genetics*, Iowa State University Press.
37. R. M. Kliman, and J. Hey. (1993). *Molec. Biol. Evol* 10, 1239–1258.
38. C. F. Aquadro, K. M. Lado, and W. A. Noon. (1988). *Genetics* 119, 875–888.
39. E. C. Kindahl, and C. F. Aquadro. (unpublished).
40. R. R. Hudson. (1982). *Genetics* 100, 711–719.

5

Effects of Genetic Recombination and Population Subdivision on Nucleotide Sequence Variation in *Drosophila ananassae*

Wolfgang Stephan

Abstract. *Genetic variation in regions of high and low recombination rates was quantified at the DNA sequence level for two natural* Drosophila ananassae *populations from Myanmar and India. Gene flow between these populations is limited. Levels of DNA sequence variation in regions of low crossing-over per physical length are significantly lower than in regions of intermediate to high crossing-over. Furthermore, fixed between-population differences were found in low crossing-over regions, but not in regions of high crossing-over. In the latter ones, frequency shifts in polymorphisms between populations are more gradual. Simple models of directional selection in conjunction with hitchhiking explain these observations only partially. Local adaptive sweeps have to be postulated to account for the rapid genetic differentiation in regions of restricted recombination.*

Introduction

To date, studies of DNA polymorphism have been carried out primarily in the dipteran species *Drosophila melanogaster* and its sibling species *D. simulans*. To assess the effects of population subdivision on nucleotide sequence variation, *D. ananassae* was used. *D. ananassae* provides an interesting comparison, in that it shows significant population substructuring. The picture that emerges from analyses of inversion polymorphism,[34] isozyme polymorphism (summarized by Johnson[13]) and DNA polymorphism[18,25,26] is that *D. ananassae* exists in many semi-isolated populations. The distribution of *D. ananassae* is largely tropical[7,22,24], with its zoogeographical center thought to be in Southeast Asia.[9] In tropical and subtropical regions of the world, *D. ananassae* is one of the most abundant species, especially in and around places of human habitation. As a domestic species, *D. ananassae* has experienced some gene flow by human activity, so that it is found now on all continents, including the Americas.[9] *D. ananassae* certainly qualifies as a polytypic species. Its widespread circumtropical distribution, especially through the scattered island groups in the Pacific Ocean, has permitted recognizable genetic differences between parts of the species to become so well developed that geographic, completely interfertile races can be distinguished.[34]

The genetics of *D. ananassae* is sufficiently well known to use this organism in population genetic studies. Rather detailed cytological and genetic maps have been worked out in the past 50 years.[32,33] Of particular interest for a comparison with *D. melanogaster and D. simulans* is the fact that the *X* chromosome of *D. ananassae* is metacentric and that several rearrangements that involve genes in the centromeric region have been fixed in these species since their last common ancestor. Genes located in the centromeric region of the *D. ananassae X* chromosome exhibit the "centromere effect",[2,20] i.e. reduced levels of recombination.[26]

I review here our analysis of DNA polymorphism at four loci of the *X* chromosome: *vermilion (v), furrowed (fw), forked (f), and Om (1D).*[25,26,27] These loci were selected such that they had different recombinational (and physical) distances from the centromere. Figure 1 shows both their cytological and genetical map positions. Comparison of the physical and genetic map suggests that crossing-over per physical length is severely repressed (Figure 1). The genetic map distance between *v* and *fw* is approximately 0.04 map units,[10] while the distance between *f* and *Om(1D)* is 0.5. The cytological position of *fw* is on the right arm

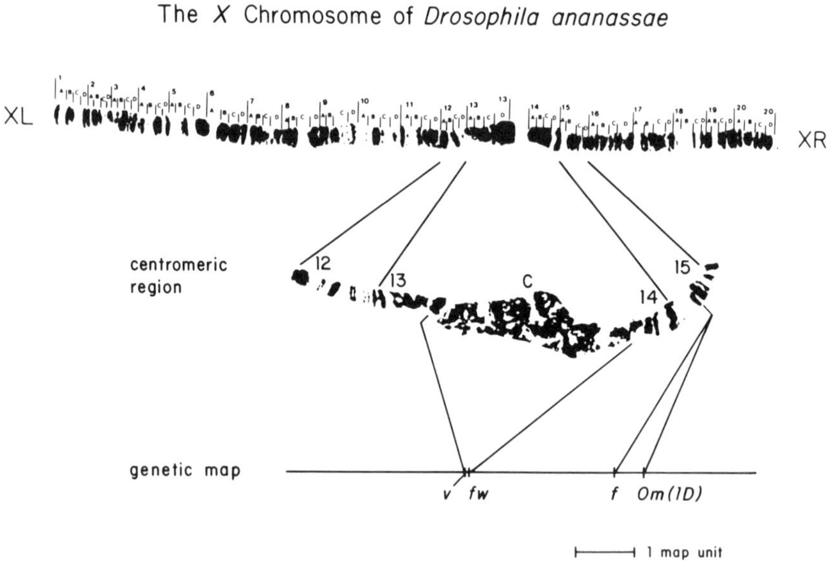

Figure 1. Physical and genetic maps of the centromeric region of the *X* chromosome of *D. ananassae*. The loci of interest in this study are indicated: *vermilion (v), furrowed (fw), forked (f),* and *Om(1D)*. Coordinates of the physical map are from Tobari, Goñi, Tomimura and Matsuda (1993).

of the *X* chromosome very close to the centromere at 14AB, i.e. in the transition zone between euchromatin and heterochromatin,[26] while *v* is located at the left arm at 13Cp.[30] *f* and *Om(1D)* map to the same polytene band. We conclude from these comparisons that the recombination frequency in the *v* locus region is very low, because the distance between *v* and *fw* spans the whole centromere, including extended regions of heterochromatin.[15] In order to examine the effects of population subdivision on variation, we chose two populations with different geographical distances from the species center in Southeast Asia: a population collected around Mandalay (Myanmar) which is presumably closest to the species center and a population from Hyderabad (India).

Patterns of Nucleotide Sequence Variation

DNA sequence variation was quantified using six-cutter restriction enzymes. The results are shown in Table 1 (average nucleotide diversity) and in Figure 2 (frequency spectra). The samples of all four loci were identical: 19 lines from the Myanmar population and 20 from the India population. The total number of restriction sites sampled for each locus are given in the inserts of Figure 2.

The *furrowed* region: In this region, we did not find a restriction site polymorphism within a population; i.e., $\pi = \theta = 0$. However, two of the 45 sites scored occurred as fixed differences between the India and Myanmar populations.[27] Fixed differences are defined as sites at which all of the sequences in one sample are different from all of the sequences in a second sample. Only one large insertion/deletion (approximately 2 kb) was found, occurring also as a fixed difference; i.e., it was present in all lines from the Myanmar population, but absent in the India population sample.

The *vermilion* region: Somewhat higher values of average nucleotide diversity were found in the *v* region (Table 1) because several restriction sites were polymorphic. The frequency spectrum of polymorphic sites (Figure 2) shows no fixed differences between both populations. One polymorphism (rank #1) is nearly fixed between both populations, the others are in low frequency. Because of this excess of rare polymorphisms, Tajima's[29] D statistic which measures the difference between π and θ should give negative values. Indeed, we obtained $D = -1.51$ and -0.77 for the Myanmar and India populations, respectively. Both values were not significantly different from zero, presumably because the number of segregating sites was too small to allow for a powerful test.

The *Om(1D)* and *forked* regions: Levels of nucleotide diversity in the *Om(1D)* and *f* regions are much higher than at *fw* or *v* (Table 1). Using the known distribution of segregating sites under a neutral infinite-site model with zero-recombination,[31] it can be shown that the observed differences in θ between the low-recombination loci *fw* and *v* and the high-recombination regions *Om(1D)* and *f* are statistically significant.[27] In other words, these heterogeneity tests reveal that the estimated θ values are not identical among all four loci.

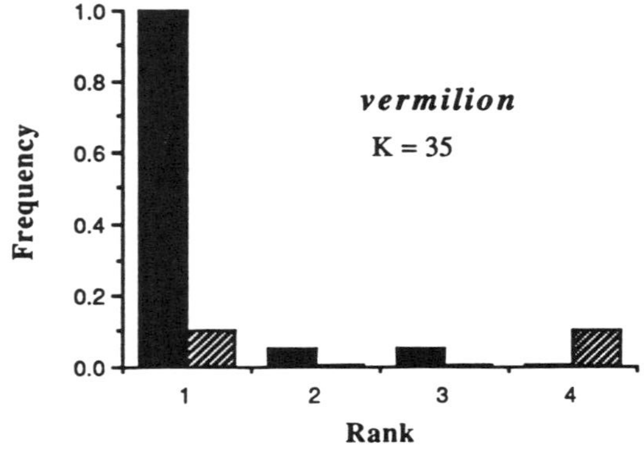

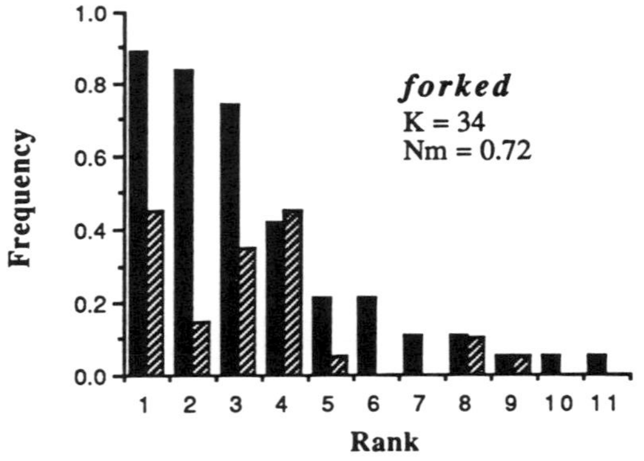

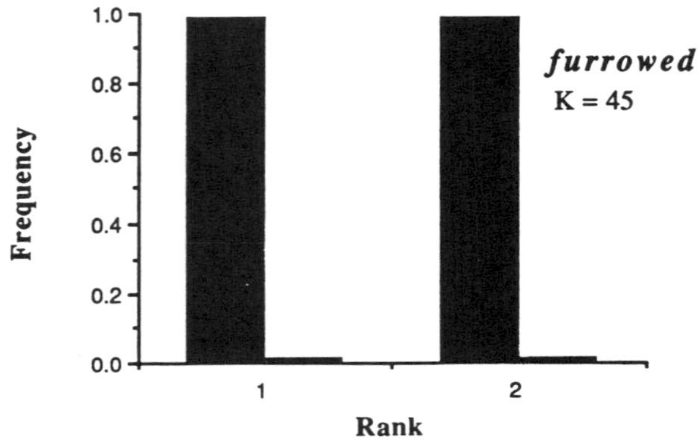

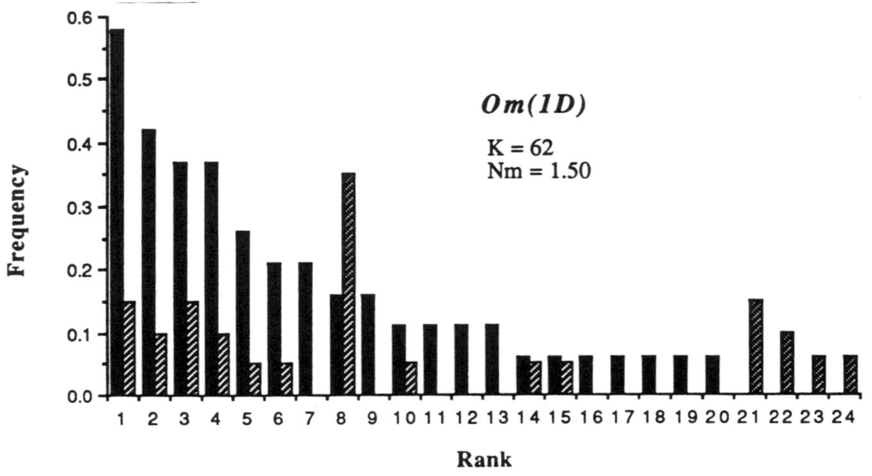

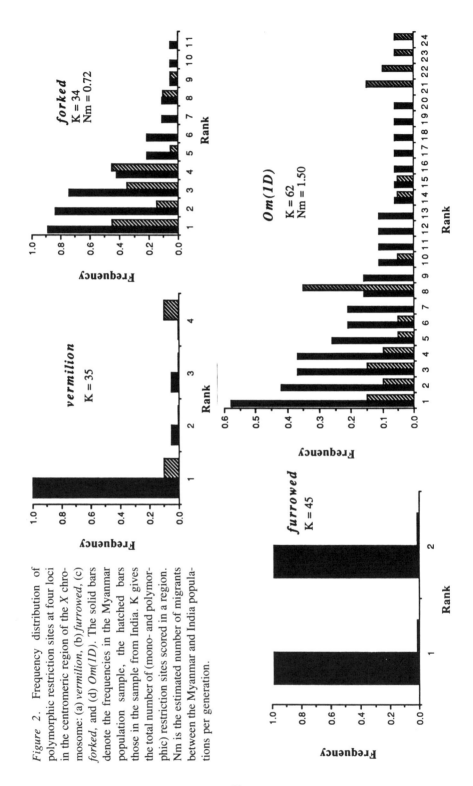

Figure 2. Frequency distribution of polymorphic restriction sites at four loci in the centromeric region of the *X* chromosome: (a) *vermilion*, (b) *furrowed*, (c) *forked*, and (d) *Om(1D)*. The solid bars denote the frequencies in the Myanmar population sample, the hatched bars those in the sample from India. K gives the total number of (mono- and polymorphic) restriction sites scored in a region. Nm is the estimated number of migrants between the Myanmar and India populations per generation.

Table 1. Estimates of nucleotide variation in the centromeric region of the X *chromosome in two* D. ananassae *populations*

		Myanmar	India
vermilion	π	0.0005	0.0009
	θ	0.0017	0.0016
furrowed	π	0	0
	θ	0	0
forked	π	0.0083	0.0062
	θ	0.0121	0.0066
Om(1D)	π	0.0085	0.0043
	θ	0.0102	0.0080

θ and π were estimated using Hudson's (1982) and Nei and Li's (1979) measures of nucleotide diversity.

Discussion

Patterns of variation appear to correlate with recombination rate: Our six-cutter survey of the *fw* and *v* regions in two *D. ananassae* populations found three salient results. 1) Average nucleotide diversity in these proximal gene regions was low within each population relative to the more distal regions *forked* and *Om(1D)*. In the *fw* region, no DNA polymorphism was detected within each population. 2) The frequency spectrum of restriction site polymorphisms in the *v* region was skewed in both populations. The frequency of site polymorphisms in each sample was low, in particular, in the sample from Myanmar. Tajima's D, a measure of skewness, was negative, although the latter result was not statistically significant. 3) The distribution of DNA polymorphisms between the two populations shows an interesting pattern. At *fw*, three fixed differences between the Burma and India populations were detected (two restriction site differences and one insertion/deletion (approximately 2 kb) difference). At *v*, all three DNA polymorphisms with high frequencies in the (total) sample (*Ins (f)*, *Ins(h)−(l)*, and the *Bam*HI-site at position 10.6; see Stephan and Mitchell[27]) occurred as nearly fixed differences, although 100% fixations were not detected. This pattern is completely different from the distribution of polymorphisms at *forked* and *Om(1D)*. Although the frequencies of several polymorphisms at the latter two loci are significantly different between the two populations, only a small minority of the differences is nearly fixed between populations, and none was completely fixed. Most polymorphisms were in intermediate frequencies in both populations (see Figure 2).

This pattern of within-population variation and between-population differentiation appears to correlate with recombination rate. Cytological studies,[26,30] combined with genetical mapping,[10] indicate that recombination rates in *v* and *fw* are much lower than in *forked* and *Om(1D)*. Furthermore, it is likely that recombination rate in the *fw* gene region is lower than in the *v* region. *fw* maps to the junction between centric heterochromatin and euchromatin on the right *X* chromosome arm, where no clear polytene banding pattern exists, whereas *v* is located in a region of the left arm, where polytene bands are still clearly visible. In fact, *v* maps to the last major band, but there are two more minor bands recognizable between this band and the centromere.[30] This evidence suggests that recombination rate is lowest in the *fw* region, increases in the *v* region and is much higher in *forked* and *Om(1D)*. In parallel, our population surveys indicate an increase of levels of within-population variation from *fw* to *v*, and to *forked* and *Om(1D)*. Fixed differences between populations disappear in that same order and are replaced by a more gradual pattern of molecular evolution. Rather than observing a distribution of fixations, a broad range of frequencies of polymorphisms is found (Figure 2).

A similar correlation between rates of crossing-over per physical length and levels of within-population nucleotide diversity has been found in *D. melanogaster*[1,4,6,17,19] and *D. simulans*.[3,6,17,19] However, fixed between-population differences have not been detected in these two species. A recent study of African and North American populations of *D. melanogaster* by Begun and Aquadro[5] comes closest to our observation of fixed differences between *D. ananassae* subpopulations.

The between-population data rule out neutral explanations: The θ-heterogeneity tests indicate that average nucleotide heterozygosity at *v* and *fw* is low relative to the levels of polymorphism at *f* and *Om(1D)*. There are several explanations for reduced variation in this gene region. It may be explained by the neutral theory[16] by assuming a reduced mutation rate. When only intraspecific data are available, this hypothesis could be tested by the method of Tajima,[29] at least in principle. However, because too few within-population polymorphisms were detected in the *v* region and none at *fw*, the Tajima test has no power or gives a trivial result. However, between-population data can be used to rule out neutral explanations. I will outline this argument here for the *fw* locus. We have observed that all sites segregating in the total sample of 39 chromosomes from Myanmar and India are fixed in one population or the other. Whether this result deviates significantly from predictions of the neutral theory depends on the amount of gene flow between these two populations. A measure of gene flow between these populations can be obtained in an island model[36] at equilibrium from the polymorphism data of the *f* and *Om(1D)* regions, since the pattern of DNA polymorphism at these two loci appears to be neutral.[25,26] Using Weir and Cockerham's[35] method for estimating F_{ST} and treating individual restriction sites as separate loci,[12] we find the following estimates of Nm: 0.72 for *f* and 1.5 for *Om(1D)*. This result indicates that the rate of migration between the Myanmar and India populations is restricted, in agreement with the conclusions derived from chromosomal and allozyme data. The following argument shows that fixed differences can not be expected under a neutral model, when migration rate is so low.

We consider a two-population symmetric island model. Each subpopulation consists of N diploids. Each generation, a small fraction m of each subpopulation is made up of migrants from the other subpopulation. We examine the genealogical history of a sample of n_1 genes from subpopulation 1 and n_2 genes from subpopulation 2. A sufficient condition for the appearance of fixed differences is that genes sampled within a subpopulation coalesce within the same subpopulation (without migrating to the other subpopulation) and that mutations occur on the branches connecting the most recent common ancestor (MRCA) of the n_1 genes in subpopulation 1 with that of the n_2 genes of subpopulation 2. This condition is also necessary, if we neglect higher-order effects which result by multiple migration events of a lineage, such as emigration and return. Since $m \sim N^{-1}$, we can neglect second- and higher-order terms. We ask whether the first part of this condition that the lineages of a sample of genes can be traced back to a MRCA within the same subpopulation is consistent with the amount of gene flow between the Myanmar and India populations which we estimated from the *f* and *Om(1D)* data ("heterogeneity of 4Nm"). Let P be the probability that each individual in the genealogy of a subpopulation had a resident as parent. Then, $P = B_0(T,m)$, where B denotes the binomial distribution and $T = T_1 + T_2$ is the sum of the total times of the genealogies of subpopulations 1 and 2. Thus, we have $P = (1-m)^T < e^{-mT}$. Since the total time of genealogy is likely to be greater than 4N for samples of size 19 or 20, we have $P < e^{-8Nm}$. For Nm = 0.72, the right-hand side is 0.0033. Hence, our estimate of Nm = 0.72 for *forked* is inconsistent with our finding of fixed differences at *fw*.

Another neutral explanation would be that the effective sizes of *D. ananassae* populations are small. However, this explanation can be ruled out because small population sizes should affect genes in low and high recombination regions in similar ways.

Do selectionist explanations work? There are currently two viable selective explanations for reduced variation in regions of low recombination. First, background selection against deleterious alleles maintained by mutation may cause the reduction in the amount of genetic variability at linked neutral sites.[8] A problem with this model is that it does not predict a large negative value of Tajima's D. For large populations such as Drosophila, the model of Charlesworth *et al.* predicts a D value of about 0. This is because a neutral variant can only remain in a large population for a sufficiently long time (and hence contribute to nucleotide diversity) if it is maintained in gametes that are free of deleterious alleles. That means that genetic variation is essentially determined by a neutral model, such that effective population size is reduced to the mutation-free allelic class. Although our result was not significant, the D values we found at *v* do not seem to be consistent with this prediction. However more data are needed to put this conclusion on a firm basis. Our Tajima tests for each population were based on only two segregating sites.

A model of genetic hitchhiking[14,21,28] associated with directional selection can explain the observed reduction in genetic diversity in regions of low recombination and also the negative D values. But there are problems with the observation of fixed genetic differences between populations at *fw*. Given the amount of gene flow estimated for *f* and *Om(1D)*, one should expect that the selected allele after completing a sweep in one population migrates to the other one. If this results in a sweep in the second population, we should not find fixed differences. Instead, hitchhiking should lead to greater homogeneity between subpopulations. However, the occurrence of fixed differences can be explained, if we assume two independent selective sweeps, driven by different *locally* adapted alleles.

This explanation raises several crucial questions. How frequent do selective sweeps occur in a population? What causes alleles to be strongly selected in one population, but not in the other? These questions are hard to answer. On the theoretical side, this requires the construction of more complex models which go beyond the simple one-locus Fisherian view of evolution assumed in the current hitchhiking models. On the empirical side, there is a need to obtain a more detailed picture of the pattern of variation in the centromere region of the *X* chromosome in *D. ananassae* populations. The biogeography of *D. ananassae* with its many genetically distinct races and the DNA polymorphism data presented here indicate that a lot more can be learned from this species on the interaction between population ecology and genetics.

References

1. M. Aguadé, N. Miyashita and C. H. Langley. (1989). Reduced variation in the *yellow-achaete-scute* region in natural populations of *Drosophila melanogaster*. *Genetics* **122,** 607–615.
2. G. W. Beadle. (1932). A possible influence of spindle fibre on crossing over in Drosophila. *Proc. Natl. Acad. Sci.* USA **18,** 160–165.
3. D. J. Begun, and C. F. Aquadro. (1991). Molecular population genetics of the distal portion of the *X* chromosome in Drosophila: Evidence for genetic hitchhiking of the *yellow-achaete* region. *Genetics* **129,** 1147–1158.
4. D. J. Begun, and C. F. Aquadro. (1992). Levels of naturally occurring DNA polymorphism correlate with recombination rates in *D. melanogaster*. *Nature* **356,** 519–520.
5. D. J. Begun, and C. F. Aquadro. (1993). African and North American populations of *Drosophila melanogaster* are very different at the DNA level. *Nature* **365,** 548–550.
6. A. J. Berry, J. W. Ajioka and M. Kreitman. (1991). Lack of polymorphism on the Drosophila fourth chromosome resulting from selection. *Genetics* **129,** 1111–1117.
7. I. R. Bock, and M. R. Wheeler. (1972). The *Drosophila melanogaster* species group. Univ. Texas. Publ. **7213,** 1–102.
8. B. Charlesworth, M. T. Morgan and D. Charlesworth. (1993). The effect of deleterious mutations on neutral molecular variation. *Genetics* **134,** 1289–1303.

9. T. Dobzhansky, and A. Dreyfus. (1943). Chromosomal aberrations in Brazilian *Drosophila ananassae*. *Proc. Natl. Acad. Sci. USA* **29,** 301–305.
10. C. W. Hinton. (1988). Formal relations between *Om* mutants and their suppressors in *Drosophila ananassae*. *Genetics* **120,** 1035–1042.
11. R. R. Hudson. (1982). Estimating genetic variability with restriction endonucleases. *Genetics* **100,** 711–719.
12. R. R. Hudson, M. Slatkin and W. P. Maddison. (1992). Estimation of levels of gene flow from DNA sequence data. *Genetics* **132,** 583–589.
13. F. M. Johnson. (1971). Isozyme polymorphisms in *Drosophila ananassae:* Genetic diversity among island populations in the South Pacific. *Genetics* **68,** 77–95.
14. N. L. Kaplan, R. R. Hudson and C. H. Langley. (1989). The "hitchhiking effect" revisited. *Genetics* **123,** 887–899.
15. H. Kikkawa. (1938). Studies on the genetics and cytology of *Drosophila ananassae*. *Genetics* **20,** 458–516.
16. M. Kimura. (1983). *The Neutral Theory of Molecular Evolution*. Cambridge University Press, Cambridge (England).
17. C. H. Langley, J. MacDonald, N. Miyashita and M. Aguadé. (1993). Lack of correlation between interspecific divergence and intraspecific polymorphism at the *suppressor of forked* region in *Drosophila melanogaster* and *Drosophila simulans*. *Proc. Natl. Acad. Sci. USA* **90,** 1800–1803.
18. M. Lynch, and T. J. Crease. (1990). The analysis of population survey data on DNA sequence variation. *Mol. Biol. Evol.* **7,** 377–394.
19. J. M. Martín-Campos, J. M. Comerón, N. Miyashita and M. Aguadé. (1992). Intraspecific and interspecific variation at the *y-ac-sc* region of *Drosophila simulans* and *Drosophila melanogaster*. *Genetics* **130,** 805–816.
20. K. Mather. (1939). Crossing over and heterochromatin in chromosomes of *Drosophila melanogaster*. *Genetics* **24,** 413–435.
21. J. Maynard Smith, and J. Haigh. (1974). The hitchhiking effect of a favorable gene. *Genet. Res.* **23,** 23–35.
22. D. Moriwaki, and Y. N. Tobari. (1975). *Drosophila ananassae,* pp. 513–535 in *Handbook of Genetics,* Vol. 3, edited by R. C. King, Plenum, New York.
23. M. Nei, and W.-H. Li. (1979). Mathematical model for studying genetic variation in terms of restriction endonucleases. *Proc. Natl. Acad. Sci. USA* **76,** 5269–5273.
24. J. T. Patterson, and W. S. Stone. (1952). *Evolution in the Genus Drosophila*. Macmillan, New York.
25. W. Stephan. (1989). Molecular genetic variation in the centromeric region of the *X* chromosome in three *Drosophila ananassae* populations. II. The *Om(1D)* Locus. *Mol. Biol. Evol.* **6,** 624–635.
26. W. Stephan, and C. H. Langley. (1989). Molecular genetic variation in the centromeric region of the *X* chromosome in three *Drosophila ananassae* populations. I. Contrasts between the *vermilion* and *forked* loci. *Genetics* **121,** 89–99.
27. W. Stephan, and S. J. Mitchell. (1992). Reduced levels of DNA polymorphism

and fixed between-population differences in the centromeric region of *Drosophila ananassae*. *Genetics* **132,** 1039–1045.

28. W. Stephan, T. H. E. Wiehe and M. W. Lenz. (1992). The effect of strongly selected substitutions on neutral polymorphism: Analytical results based on diffusion theory. *Theoret. Pop. Biol.* **41,** 237–254.
29. F. Tajima. (1989). Statistical method for testing the neutral mutation hypothesis by DNA polymorphism. *Genetics* **123,** 585–595.
30. S. Tanda, A. E. Shrimpton, C. W. Hinton and C. H. Langley. (1989). Analysis of the *Om(1D)* locus in *Drosophila ananassae*. *Genetics* **123,** 495–502.
31. S. Tavaré. (1984). Line-of-descent and genealogical processes and their applications in populations genetics models. *Theoret. Pop. Biol.* **26,** 119–164.
32. Y. N. Tobari. (1993). Linkage maps. In: *Drosophila ananassae—Genetical and biological aspects* (Y. N. Tobari, ed.). Japan Scientific Societies Press, Tokyo, and Karger, Basel.
33. Y. N. Tobari, B. Goñi, Y. Tomimura and M. Matsuda. (1993). Chromosomes. In: *Drosophila ananassae—Genetical and biological aspects* (Y. N. Tobari, ed.). Japan Scientific Societies Press, Tokyo, and Karger, Basel.
34. Y. Tomimura, M. Matsuda and Y. N. Tobari. (1993). Polytene chromosome variations of *Drosophila ananassae* and its relatives. In: *Drosophila ananassae—Genetical and biological aspects* (Y. N. Tobari, ed.). Japan Scientific Societies Press, Tokyo, and Karger, Basel.
35. B. S. Weir, and C. C. Cockerham. (1984). Estimating *F*-statistics for the analysis of population structure. *Evolution* **38,** 1358–1370.
36. S. Wright. (1951). The genetical structure of populations. *Ann. Eugenics* **15,** 323–354.

6

Polymorphism and Divergence in Regions of Low Recombination in *Drosophila*

Montserrat Aguadé and Charles H. Langley

Introduction

Surveys of DNA sequence variation within Drosophila populations were initially aimed at describing standing levels of substitutional and insertion-deletion polymorphism in different regions and in different species. While the general goal was a broad picture of variation, the next goal was to try to infer the mechanisms responsible for patterns of variation. For example, variation at loci on the sex chromosomes was found to be comparable to that at autosomal loci, i. e. no evidence for any reduction of variation was seen. While there is little power in these studies to test the expected ¾ reduction in heterozygosity expected under pure selective neutrality, they do rule out recessive, mildly deleterious mutations as a major component of DNA sequence polymorphisms. In surveys of restriction-map variation at the *white*[28] and *zeste-tko*[1] regions in 64 lines of *D. melanogaster*, the level of variation attributable to nucleotide substitutions or the estimated heterozygosity per nucleotide (0.009 and 0.004, respectively) was similar to that previously reported for autosomal regions. These estimates provided no support for models that predicted less polymorphism on the X chromosome.

A second and more powerful approach in the study of variation within populations is to compare it with divergence between species. According to the neutral theory of molecular evolution[19] there is a quantitative relationship between polymorphism and divergence in a given region of the genome. Different regions can be compared quantitatively for consistency. Selection at linked sites can, however, distort the level of neutral polymorphism in a given region causing either a reduction (e. g. directional selection) or an increase (e. g. balancing selection) with respect to neutral expectations. The quantitative predictions of relative numbers of segregating sites (and divergent sites) in different regions of the genome is a straightforward extention of the neutral theory.[15] The reduction in polymorphism due to directional selection at linked loci depends on both the intensity of selection and linkage (actually the ratio of these two) and the density (in time and per unit map distance) of selected substitutions.[18,27] Figure 1a shows the characteristic impact of such linked directional selection on the phylogeny of sample allelic DNA sequences. A linked balanced polymorphism will lead to an increase in the number of segregating sites in a region closely linked to the selected site.[15,16] Figure 1c diagrams the expected consequences of a linked balanced polymorphism (compared to Figure 1b). Although adequate estimates of the level of polymorphism in a given region can be obtained both by restriction-map analysis and by sequencing, estimates of divergence are best obtained by direct sequencing. The number of regions to which such divergence and polymorphism comparisons have been applied is limited. Nevertheless the results have been remarkable in that many clear instances of deviations from the predictions of the neutral theory have been reported. In the initial application of the HKA test[15] comparing silent site polymorphism and divergence in the *Adh* and its 5′ flanking region, an excess of polymorphism was detected at the *Adh* coding region compatible with the existence of a balanced polymorphism.

Under the directional hitchhiking effect model selectively favored mutations (or rare variants) in a region of restricted crossing over will sweep through the population. Variation after such a selective sweep will be zero in the region of the genome surrounding the selected locus. With time, selectively neutral mutation will slowly introduce new polymorphic sites, but both the number of neutral polymorphisms and their frequencies are expected to be reduced for many generations. But such selective sweeps are not expected to reduce the level of interspecies divergence. This hitchhiking scenario was proposed to explain the consistent and striking observation of reduced polymorphism (relative to other genomic regions) in genomic regions of low crossing over per physical length.[18] Below we review some of these recent studies.

Analysis of Variation at Regions with No Measurable Crossing Over per Physical Unit Length

In *D. melanogaster* there is a contraction of the genetic maps both in telomeres and centromeres; on the *X* chromosome this effect is marked and well documented.[23] Both the *yellow* gene and the *ASC* region had been cloned and characterized (Campuzano *et al.* 1985) offering an opportunity to study variation at the DNA level at those regions. Our first survey of restriction-map variation at the *y - ASC* region in 64 lines of *D. melanogaster* showed reduced nucleotide variation:[2] only nine polymorphisms were detected out of 176 sites scored (approximately 2,112 nucleotide site equivalents). The frequency spectrum of those polymorphisms was skewed with an excess of rare polymorphisms: six out of the nine polymorphisms were present only once in the sample. Other authors,

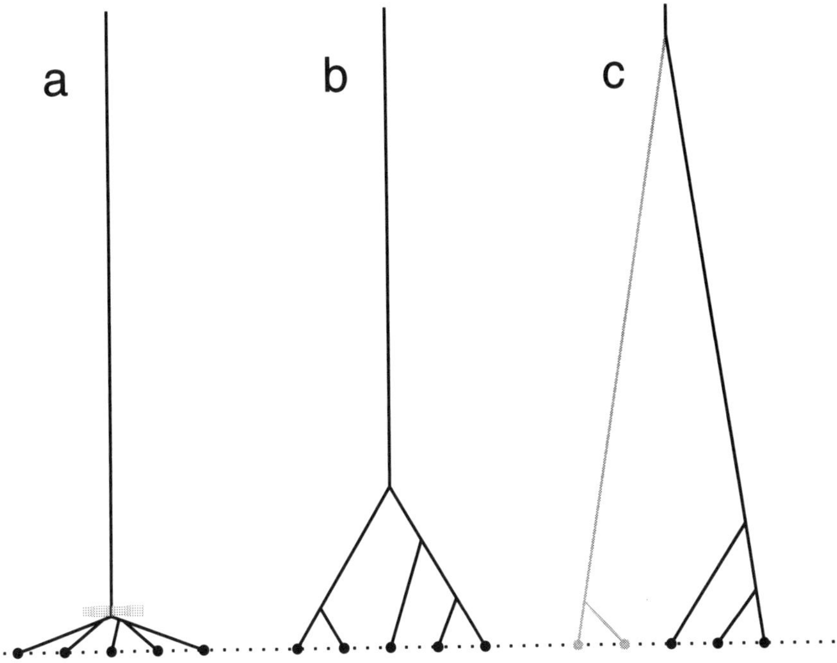

Figure 1. Possible gene genealogies of a random sample of five alleles. **a**. depicts the impact of a recent directional hitchhiking event in the history of the alleles in the sample. **b**. depicts a *typical* history under random drift. **c**. depicts the impact of a closely linked balanced polymorphism. The mean time to the common ancestor of any two alleles is expected to be 2N generations (N is the population size) in the absence of selection, tree **b**. The time to the common ancestor of the whole sample will be the time back to the most recent selective sweep of a closely linked favored rare variant, the gray stripe in tree **a**. The time back to the common ancestor of the alleles closely linked to two alleles of a balanced polymorphism might be as long as the time the polymorphism arose, the *gray* line of descent being linked to one selected allele and the *black* to the other. Thus the total time in the genealogy of alleles near a recent directional hitch hiking event will be much less than that in a purely neutral genealogy. On the other hand the time in the genealogy of alleles closely linked to a balanced polymorphism can be much greater than that of a purely neutral genealogy. Thus the numbers of selectively neutral segregating (polymorphic) sites will be fewer for the first and greater for the last. Also notice that most mutations that might fall on tree **a** would be unique (1:4) while most on tree **c** would not be unique (2:3). This difference in the frequency spectra of segregating sites will lead to differences in the statistic *D* proposed by Tajima.

however, studying this same region could not find any evidence of reduced variation in their own data[5,13,25] and our interpretation of reduced variation was challenged. W. Stephan (pers. communication) found evidence of excess of unique polymorphisms in our sample. Later Hudson[14] confirmed that given the sample size and the number of sites scored, the probability of detecting nine or fewer polymorphisms given the average heterozygosity per nucleotide in Drosophila (0.005) was negligible (2×10^{-6}). A population bottleneck effect could be discarded because as previously described the same sample had also been analyzed for other regions of the *X* chromosome (*w* and *z-tko*) with no evidence for any reduction in levels of variation.

However, only interspecific comparison in this region could eliminate the possibility of differential constraint, i. e. a reduction in the rate of mutation to selectively neutral variants in regions of reduced crossing over. When divergence at the *achaete* region between *D. melanogaster* and *D. simulans* was estimated by sequence comparison,[26] no evidence for any reduction in interspecies divergence relative to other regions was found (0.07 and 0.06 for silent and for all sites at the *ac* region) (Table 1). This lack of reduction in interspecies divergence indicated that there was no differential constraint or mutation in this region of the tip of the *X* chromosome, and therefore discarded purifying selection as the mechanism responsible for the reduced level of polymorphism in this region of *D. melanogaster*.

Nucleotide variation at the *achaete* region was also studied in samples from natural populations of the sibling species, *D. simulans*.[26] An even greater reduction in polymorphism within this species was found; no nucleotide polymorphism was detected in a sample of 103 lines when studying 390 site equivalents (Table 2). The 95% upper bound on the estimated values of $3N_e\mu$ ($3N_e\mu$ is equal to the expected heterozygosity per nucleotide under selective neutrality for an X-linked locus) compatible with the observation of no polymorphism[14] is 0.002, an order of magnitude lower than the estimated value for the *rosy* region (0.018).[4] Given that the *yellow-achaete* region in *D. simulans* is also located at the tip of the *X* chromosome, this reduction of variation is most easily explained by the hitchhiking effect of the fixation of a linked favorable selected substitution in this species.

Table 1. Estimates of silent site divergence between D. melanogaster *and* D. simulans *at different genomic regions*

	Silent Site Divergence
ac	0.070
su(f)	0.120
5'Adh	0.068
hsp82	0.057
Mtn	0.075

Table 2. Summary of nucleotide variation as revealed by four-cutter analysis at the y-ASC *and* su(f) *regions*

		Nucleotide Diversity	
		D. melanogaster	*D. simulans*
y-ASC		0.0004 (n=245)	0.000 (n=103)
	Europe	0.0002 (n=190)	
	North America	0.0012 (n=47)	
su(f)		0.0002 (n=55)	0.0005 (n=103)

In order to corroborate our original conclusion of reduced variation at the *y-ASC* region in *D. melanogaster*,[2] we surveyed restriction-map variation by four-cutter analysis at all major transcription units of that region in 10 populations of this species.[26] Variation was actually reduced throughout this region (Table 2). Application of the HKA test to silent site variation at this telomeric region in contrast to that at the *Adh* 5′ flanking region confirmed that this reduction of variation was not due to differential selective constraint at the tip of the *X* chromosome. Some population differentiation between European and North American samples was evident. Although European samples showed an excess of rarer variants, no skewness in the frequency spectrum was detected in the North American samples. Deviations in the frequency spectrum from neutral expectations can be tested using Tajima's statistic D.[32] This statistic measures the relative difference between two different estimators of $4N_e\mu$ ($3N\mu$ for sex-linked loci), θ and π (nucleotide diversity); θ only takes into account the number of polymorphisms in a given region but not their frequencies, while π (essentially expected heterozygosity) does. In a mutation-drift equilibrium D should be very close to zero as both θ and π should be estimating the same thing. Skewness in the frequency spectrum toward excess of rare variants should result in significant negative values of D, as is the case in European populations. Negative values of D can be also due to recent bottlenecks. Like in our previous survey by six-cutter analysis,[2] bottleneck effects in European populations could be discarded given that the largest sample (that from Barcelona) was also analyzed for the *white* region and there was no evidence for any reduction in the level of nucleotide variation.[29]

Su(f) is located in the centromere-proximal region of the *X* chromosome where crossing over per physical unit is also reduced. It was important to know whether the observations of reduced variation at the tip of the *X* chromosome[2,6,26] and in the fourth chromosome[9] could also be extended to this other region of restricted recombination. We have therefore analyzed variation both within *D. melanogaster* (55 lines) and *D. simulans* (103 lines) for the *su(f)* region.[22] Similar to our results for the *y-ASC* region there is a reduction of nucleotide heterozygosity both in *D. melanogaster* and *D. simulans* (Table 2). As in the case of the *y-ASC* region we have estimated divergence by sequence comparison and found no evidence for any reduction in interspecies divergence at this region; the estimated divergence (0.12 substitutions per silent site) is actually at the upper end of the range of the estimates reported for several gene regions between these two species (Table 1). Application of the HKA test comparing this region to the *Adh* 5′ flanking region revealed a significant deviation from neutral expectations. Again the lack of correlation between levels of polymorphism and interspecific divergence is best explained by selective sweeps both in *D. melanogaster* and *D. simulans* reducing intraspecific variation.

The general picture emerging from the studies both in *D. melanogaster* and *D. simulans* in regions with no measurable crossing over (the most distal tip and most centromere-proximal base of the *X* chromosome, and fourth chromosome) is: 1) at the intraspecific level, reduced heterozygosity, skewed frequency spectrum in some populations (excess of rare polymorphisms) and some population differentiation, and 2) at the interspecific level, no reduction in levels of divergence.

Although both the strong reduction in heterozygosity and the lack of reduction in divergence can be best explained by the hitchhiking model, some of the other observations (lack of skewness in the frequency spectrum in some populations and population differentiation) are not fully consistent with the theory at present.

Analysis of Variation at Regions with Reduced but Measurable Crossing Over per Physical Unit Length

As reported both in *D. melanogaster* and in *D. simulans* surveys of DNA variation in regions of low crossing over per physical unit length have concentrated in regions with no measurable crossing over. Analysis of variation in regions with increasing and measurable levels of crossing over may allow more quantitave tests of the correlation between levels of variation and levels of crossing over predicted by the hitchhiking effect model and suggested by the analyses of Begun and Aquadro[7] and Wiehe and Stephan.[33] As depicted in Figure 2 the tip of the *X* chromosome of *D. melanogaster* shows a monotonic increase of levels of crossing over per physical unit length from the telomere inward. This region seems specially suitable for such studies as many extensive regions have been cloned (by chromosome walking) and are partially or totally sequenced. On the other hand there are many genetic tools available in this region to obtain good estimates of rates of crossing over.[24]

To test more quantitative predictions of the hitchhiking model or other alternative models, specifically the expected skewness in the site frequency spectrum and the expected increased level of linkage disequilibrium, we recognize that

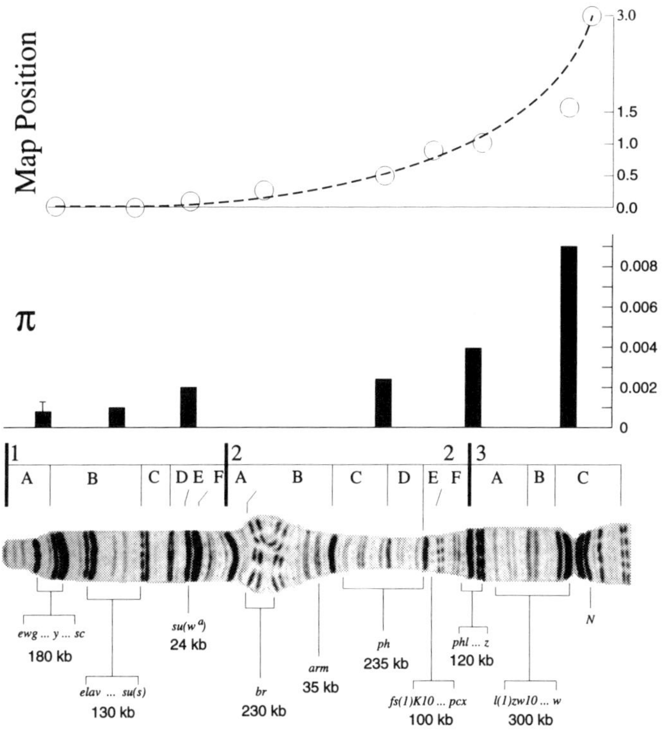

Figure 2. Relationship between the genetic and cytological maps for different telomere-proximal regions of the X chromosome of *D. melanogaster*. The map distance of various loci to *yellow* is plotted against cytological position in the upper graph. The lower graph shows the average number of differences per nucleotide (π) for several studied regions is also plotted: *y-ASC* (Martín-Campos *et al.* 1992), *su(s)* and *su(w*^{a}) (Aguadé et al. 1994), *Pgd* (Begun and Aquadro 1993) and *z-tko* (Aguadé et al. 1989a), *white* (Miyashita and Langley 1988). Regions cloned (generally by chromosome walking) and characterized have been indicated below the figure of the polytene chromosome.

we should be able to survey larger segments for polymorphism in a given region and to survey larger sample sizes. Although sequencing would seem the method of choice because it detects all nucleotide polymorphisms in a given region, it does not seem adequate to survey large regions (given the expected low density of polymorphisms) nor large sample sizes. SSCP (Single Strand Conformation Polymorphism)[30] coupled with stratified sequencing seemed promising for our goal. Once SSCP classes for each studied fragment are scored, two (or more) alleles must be sequenced per class to assess any intraclass variation and to determine the sequence differences responsible for the class differences. In our first application we have tested the techniques efficiency and reliability in detecting DNA sequence variation in regions of low polymorphism.[3]

Our first effort to study variation at regions with increasing and measurable levels of crossing over has focused on the *su(s)* and the *su(wa)* regions.[3] Figure 2 shows the positions of the *su(s)* and the *su(wa)* (as well as *y* - *ASC*) on the tip of the *X* chromosome of *D. melanogaster*. The application of SSCP with stratified DNA sequencing to a sample of 50 chromosomes from a North American natural population yielded 10 substitutional polymorphisms and three insertional polymorphisms in 3,213 bp of surveyed noncoding sequence in the *su(s)* region and 17 substitutional polymorphism and six insertional polymorphisms in 1,955 bp of surveyed noncoding sequence in the *su(wa)* region. From the distribution of these polymorphisms it was possible to estimate the average number of substitutional differences per site at each of the loci. Table 3 shows these estimates with their bootstrapped confidence intervals; it also shows the 95% upper confidence intervals on the estimated values of $3N_e\mu$ compatible with the observations of 10 and 17 nucleotide polymorphisms in a sample of 50 alleles, respectively. These upper limits are well below the accepted average heterozygosity per nucleotide in this species, indicating a reduction in levels of DNA variation in these two regions also.

In Figure 2 the estimated average number of differences per site is also plotted for these data from *su(s)* and *su(wa)* and other loci. There is a clear pattern of correlation between the amount of crossing over (with *y*) and the observed level of variation. This result fits in a qualitative way to the predictions of the hitchhiking model (discussion elsewhere in this volume). But other properties of these data do not. As mentioned above and discussed elsewhere in this volume such hitchhiking effect should also skew the frequencies of those polymorphisms that are observed. But this is clearly not the case. In our results from *su(s)* and *su(wa)* and in those from other loci there is no support for this expectation. The estimated values of Tajima's D for the *su(s)* and the *su(wa)* data are 1.31 and –0.40, respectively. Simulation results (not shown) indicate that to produce the level of reduction in polymorphism observed in the present data sets the simple hitchhiking effect model would produce statistically significant negative D values. Thus the hitchhiking effect model is inconsistent with the observations. The only proposed alternative is the background deleterious selection model.[11] However, the authors find that an excessively high deleterious mutation rate is needed to explain the reported reduction of variation in those regions of low crossing over.

While the theoretical models obviously need some refinement it is expected

Table 3. Summary of nucleotide variation at the su(s) *and* su(wa) *regions*

	# Sites	# Polymorphisms	Nucleotide diversity	Upper Theta
su(s)	3,213	10	0.001 (0.00095–0.00111)	0.0015
su(wa)	1,955	17	0.0021 (0.0019–0.0024)	0.0036

that more detailed measurements of polymorphism in the tip of the *X* in different populations will form the basis for the evaluation of such new models. We anticipate that these future studies will also require more precise measurements of recombination in this region.

Acknowledgments

We acknowledge support from NSF Grant BSR–9117222 and NATO grant CRG–900651.

References

1. M. Aguadé, N. Miyashita and C. H. Langley. (1989a). *Mol. Biol. Evol.* **6,** 123–130.
2. M. Aguadé, N. Miyashita and C. H. Langley. (1989b). *Genetics* **122,** 607–615.
3. M. Aguadé, W. Meyers, A. D. Long and C. H. Langley. (1994). *Proc. Natl. Acad. Sci. USA* (in press).
4. C. F. Aquadro, K. M. Lado and W. A. Noon. (1988). *Genetics* **119,** 875–888.
5. R. N. Beech, and A. J. Leigh-Brown. (1989). *Genet.Res.* 53, 7–15.
6. D. J. Begun, and C. F. Aquadro. (1991). *Genetics* **129,** 1147–58.
7. D. J. Begun, and C. F. Aquadro. (1992). *Nature* (London) **356,** 519–520.
8. D. J. Begun, and C. F. Aquadro. (1993). *Nature* (London) **365,** 548–550.
9. A. J. Berry, J. W. Ajioka and M. Kreitman. (1991). *Genetics* **129,** 1111–1117.
10. R. K. Blackman, and M. Meselson. (1986). *J. Mol. Biol.* **188,** 499–515.
11. D. Charlesworth, M. T. Morgan and B. Charlesworth. (1993). *Genetics* **134,** 1289–1303.
12. V. H. Cohn, and G. P. Moore. (1988). *Mol. Biol. Evol.* **5,** 154–166.
13. W. F. Eanes, J. Labate and J. W. Ajioka. (1989). *Mol. Biol. Evol.* **6,** 492–502.
14. R. R. Hudson. (1990). Gene genealogies and the coalescent process. *Oxf. Surv. Evol. Biol.* **7:**1–44.
15. R. R. Hudson, M. Kreitman and M. Aguadé. (1987). *Genetics* **116,** 153–159.
16. R. R. Hudson, and N. L. Kaplan. (1988). *Genetics* **120,** 831–840.
17. T. H. Jukes, and C. R. Cantor. (1969). *Evolution of protein molecules*, pp.21–132 in *Mammalian Protein Metabolism*, edited by H. N. Munro. Academic Press, New York.
18. N. Kaplan, R. R. Hudson, and C. H. Langley. (1989). *Genetics* **123,** 887–899.
19. M. Kimura. (1983). *The Neutral Theory of Molecular Evolution*. Cambridge University Press, Cambridge.
20. M. Kreitman, and M. Aguadé. (1986). *Proc. Natl. Acad. Sci. USA* 86, 3562–3666.
21. B. W. Lange, C. H. Langley and W. Stephan. (1990). *Genetics* 126, 921–932.

22. C. H. Langley, J. MacDonald, N. Miyashita, N and M. Aguadé. (1993). *Proc. Natl. Acad. Sci. USA* **90,** 1800–1803.

23. D. L. Lindsley, and L. Sandler. (1977). *Philos. Trans. Soc. Lond. (Biol.)* **277,** 295–312.

24. D. L. Lindsley, and G. G. Zimm. (1992). *The Genome of* Drosophila melanogaster . Academic Press, San Diego.

25. J. N. Macpherson, B. S. Weir and A. J. Leigh-Brown. (1990). *Genetics* **126,** 121–129.

26. J. M. Martín-Campos, J. M. Comerón, N. Miyashita, N. and M. Aguadé. (1992). *Genetics* **130,** 805–816.

27. J. Maynard Smith, and J. Haigh. (1974). *Genet.Res.* **23,** 23–35.

28. N. Miyashita, and C. H. Langley. (1988). *Genetics* **120,** 199–212.

29. N. Miyashita, M. Aguadé, M. and C. H. Langley. (1993). *Genet. Res.* **62,** 101–109.

30. M. Orita, H. Iwahana, K. Kanazawa and T. Sekiya. (1989). *Proc. Natl. Acad. Sci. USA* **86,** 2766–2770.

31. R. K. Saiki, S. Scharf, F.Faloona, G. T. Horn, H. A. Erlich, and N. Arnheim. (1985). *Science* **230,** 1350–1354.

32. F. Tajima. (1989). *Genetics* **123,** 585–595.

33. T. H. E. Wiehe, and W. Stephan. (1993). *Mol. Biol. Evol.* **10,** 842–854.

7

Inferring Selection and Mutation from DNA Sequences: The McDonald-Kreitman Test Revisited

Stanley A. Sawyer

Abstract. *Aligned DNA sequences from the same species, and from two species that share a sufficiently recent common ancestor, are analyzed for evidence of unidirectional selection. We develop a technique for estimating the direction and magnitude of unidirectional selection that can be used at synonymous sites as well as at replacement codon positions. The model is applied to sample configurations at the ADH locus in two* Drosophila *species, and to the* gnd *locus in* Escherichia coli. *The analysis suggests positive selection for observed amino acid replacements at ADH, and is highly significant for selection against the observed amino acid replacements at* gnd. *Small amounts of negative selection are detected in synonymous sites, even after allowance is made for base-dependent mutation.*

Introduction

Experiments beginning in the middle 1960s found that many enzymes in natural populations are polymorphic.[1,2] A protracted and often heated debate began about the selective significance of these polymorphisms.[3] Some scientists felt that most of this variation is essentially selectively neutral, with at most negligible effects on fitness.[4,5] Others held that different enzymes were unlikely to be selectively equivalent, and so various forms of selection must be involved.[6,7,8,9,10] Later experiments showed there is even greater *synonymous* or *silent* variation at the DNA level. That is, there is more variation in the DNA itself than in the protein products.

Also puzzling is the phenomenon of *codon bias*, which is the tendency for the different three-base codons that code for the same amino acid to occur in different frequencies.[11,12] One possible reason for codon bias may be biased mutation rates at the DNA level. That is, some of the individual bases that compose DNA may be more mutable than others, or else there may be bias in their rates of mutation to other bases. Selective forces may also favor some codons over others. Particular codons may be necessary for the proper mRNA configuration, and variable tRNA abundances may slow the translation of some synonymous codons.[11,12] However, rates of nucleotide substitution at silent codon sites are similar to those in pseudogenes,[13,14] which suggests that most silent substitutions are nearly neutral. While in general silent DNA changes appear to be under weaker selective constraints than changes that cause amino acid replacement, some silent changes may have a selective effect.

A Quantitative Model for Selection

Our main purpose here is to discuss a method for detecting and estimating selection based on an aligned set of DNA sequences. The method will be applied both to silent site variation and to amino-acid variation. The model of selection will be that all changes to a consensus or an ancestral base or amino acid are either strongly deleterious (and so will never be seen in a natural population) or else change the fitness of the host by an equal amount, equal both for changes to different bases at the same site and for changes at different sites. Mutations at different sites have multiplicative effects on fitness. In particular, this model may not appropriate for detecting balancing or disruptive selection. Of course, no statistical test can detect *arbitrary* forms of selection, since *anything* that you observe could be the result of particular selective forces.

A fringe benefit of this analysis is that you can estimate separate mutation rates for synonymous and for amino-acid replacements. Many changes to a functioning protein or enzyme are presumably lethal or nearly lethal. Amino-acid variation that is common enough to be detected in a sample of DNA sequences is presumably subject to relatively weak selection. An estimate of the ratio of the amino-acid replacement mutation rate to the silent mutation rate should give an estimate of the proportion of amino acids in a protein that can be replaced without lethal or strongly deleterious effects on the host.

A Contingency-Table Test for Unidirectional Selection

Before describing the quantitative model, we first discuss a 2×2 contingency table test that can detect differences in selection between silent and replacement sites.[15] Suppose that we have n aligned DNA sequences from a coding region. Most aligned sites will have a single most common base with at most a few sequences with different bases at that site. If most changes from this consensus base are selectively deleterious, then we would expect relatively few sequences at any particular site to be different from the consensus base. Similarly, if changes from the consensus base were advantageous, we would also expect relatively few differences from the consensus, since otherwise the consensus would be quickly driven from the population. If polymorphic replacement sites are observed to be more bunched (i.e., have fewer deviations from the consensus) than polymorphic silent sites, then one possible explanation is unidirectional selection for or against the replacements.

Specifically, given n aligned DNA sequences, we say that a site is *simply polymorphic* if $n-1$ of the sequences have one base at this site and one sequence has a second base. All other polymorphic sites are called *multiply polymorphic*.

A three-site codon position is called *regular* if the consensus bases code for an amino acid other than leucine or arginine, which is equivalent to saying that the first two positions are nondegenerate (e.g., any replacement changes the amino acid). About half of regular amino acids are fourfold degenerate at the third position, which means that any base can be substituted at the third codon site without changing the amino acid. Most of the other amino acids are twofold degenerate at the third position. Two-fold degenerate amino acids are of two types. The first type can have either of the two pyrimidines T or C at the third position, but any other change at the third site changes the amino acid. For the second type, the third base is either of the two purines A or G. There is one threefold degenerate amino acid (isoleucine), which corresponds to the three codons ATT, ATC, and ATA. Since the codon ATA is extremely rare in most natural populations, we treat isoleucine as twofold degenerate and the rare codon positions with an ATA as irregular.

Consider a 2 × 2 contingency table with the numbers of silent simply and multiply polymorphic sites at amino-acid monomorphic regular codon positions in the first row, and the numbers of simply and multiply polymorphic sites at the first and second positions of regular codons in the second row (Table 1). Regular codon positions have the potential of supplying two replacement polymorphisms, but this is rare, and historically may have been the result of two amino-acid replacements.

When all silent sites are used, the contingency table in Table 1 is highly significant. However, most replacement variation may be lethal or at least subject

Table 1. *2 × 2 tables for selection at the* gnd[a] *locus in* Escherichia coli[b]

	All silent[c]		Two-fold silent[c]	
	simple poly	multiple poly	simple poly	multiple poly
Silent (regular)[c]	60	83	27	31
Replacements (1,2 pos'n regular)	20	7	20	7
	$P = 0.003$[d]		$P = 0.021$[d]	

[a] The *gnd* locus transcribes the enzyme 6-phosphogluconate dehydrogenase.

[b] 14 strains of *E. coli* (1407bp; GenBank[16,17])

[c] See text for definitions.

[d] Two-sided Fisher exact test.

to highly deleterious selection, so that there may be at most two weakly-selected bases at any amino-acid varying site. Thus it may be fairer to compare replacement polymorphisms with twofold degenerate silent polymorphisms, which are more likely to be simply polymorphic than fourfold degenerate sites. When twofold degenerate sites are used, the contingency table in Table 1 is significant but is not highly significant.

Estimating Mutation Rates from Silent Sites

There are many different ways to estimate the amount of mutation at silent sites in an aligned set of DNA sequences (see e.g. Fu and Li, 1993[18]). The following approach[10,15] has the advantage that it automatically allows for saturation and homoplasy (i.e., parallel or repeated mutations at the same site), and can be adapted to estimate the divergence time between two species as well.

We assume that fourfold degenerate sites (for example) have mutation rates μ_T, μ_C, μ_A, and μ_G *to that base* per chromosome per generation. Thus the mutation rate does not depend on the base, but depends on the *target* base. Note that "mutations" of e.g. T to T that do not change the base are permitted, but will be taken into account before estimating the locus-wide silent mutation rate. While this model is not the most common way to model base-dependent mutation, it is consistent with some models of isochores in mammals,[14] and does lead to computable base-dependent estimates of mutation rates that do not need homoplasy corrections.

Under these conditions, the *population* frequencies of the four bases at that site will have the joint probability density

$$C_{\boldsymbol{\alpha}}\, p_T^{\alpha_T-1} p_C^{\alpha_C-1} p_A^{\alpha_A-1} p_G^{\alpha_G-1}\, dp_T dp_C dp_A dp_G \tag{1}$$

where $\alpha_T = 2N_e\mu_T$, $\alpha_C = 2N_e\mu_C, \ldots$, where N_e is the haploid effective population size and $C_{\boldsymbol{\alpha}} = C(\alpha_T, \alpha_C, \ldots)$, under the usual conditions for diffusion approximations.[10,19,20] The density in equation (1) is called a Dirichlet density. The corresponding density for pyrimidine twofold degenerate sites is the beta density $C'_{\boldsymbol{\alpha}}\, p_T^{\alpha_T-1} p_C^{\alpha_C-1} dp_T dp_C$ for $p_T + p_C = 1$, with a similar expression for the purine twofold degenerate sites.

Now assume that the site is part of an aligned sample of n DNA sequences, and consider the probability that the sample has n_T sequences with the base T at that site, n_C sequences with C, n_A with A, . . ., where $n = n_T + n_C + n_A + n_G$. This probability can be obtained by integrating the density in equation (1), and is

$$C_{\mathbf{n}} \frac{\alpha_T^{(n_T)} \alpha_C^{(n_C)} \alpha_A^{(n_A)} \alpha_G^{(n_G)}}{\alpha^{(n)}}, \qquad \alpha = \alpha_T + \alpha_C + \alpha_A + \alpha_G \tag{2}$$

where $x^{(k)} = x(x+1)\ldots(x+k-1)$ and $C_n = n!/(n_T!n_C!n_A!n_G!)$.[10,21] The corresponding probability for pyrimidine twofold degenerate sites is $C'_n\alpha_T^{(n_T)}\alpha_C^{(n_C)}/(\alpha_T+\alpha_C)^{(n)}$. The probabilities in equation (2) for fourfold degenerate sites, and the corresponding probabilities at twofold degenerate sites, can be combined to obtain maximum likelihood estimators for α_T, α_C, α_A, and α_G (Table 2).

Given equation (1), the mean frequency of the base T at fourfold degenerate sites is $E(p_T) = \alpha_T/\alpha$ for α in equation (2). Thus the expected rate of transitions $T \rightarrow C$ at fourfold degenerate sites in a genetic locus is $N_4\alpha_T\alpha_C/(2\alpha)$, where N_4 is the number of fourfold degenerate regular codon positions in the locus. (The factor of two is because μ_C is the mutation rate to the base C per N_e generations, while $\alpha_C=2N_e\mu_C$ in equation (1).) Similarly, the mutation rate at pyrimidine twofold degenerate sites in the locus is $N_{2,TC}\alpha_T\alpha_C/(\alpha_T+\alpha_C)$, where $N_{2,TC}$ is the number of pyrimidine twofold degenerate regular codon positions. These considerations lead to a formula for the locus-wide silent mutation rate μ_{sil}[10] (Table 2).

The maximum likelihood method assumes that the distributions at different silent sites can be treated as independent. Recombination and gene conversion both help to insure the independence of site distributions. Independence can be tested by computing the significance of autocorrelations of the events monomorphic/polymorphic for adjacent silent sites. The first three autocorrelations are not significant for either the *E. coli* data in Table 1 nor the two ADH data sets in Table 2. Maximum likelihood theory uses independence only for the central limit theorem for the log likelihoods for the various terms,[23] and so can tolerate some deviation from joint statistical independence.

The methods that are used in this paper assume that each sample from a species is a random sample from a panmictic population. If a sample contains some

Table 2. Maximum likelihood estimates of α_T, α_C, α_A, α_G from the probabilities (2) at silent sites

ADH[a]		*gnd*[b]	
alpha's	4-fold[c]	alpha's	4-fold[c]
$\alpha_T = 0.0080$	0.155	$\alpha_T = 0.128$	0.407
$\alpha_C = 0.0300$	0.610	$\alpha_C = 0.109$	0.288
$\alpha_A = 0.0023$	0.066	$\alpha_A = 0.063$	0.106
$\alpha_G = 0.0097$	0.169	$\alpha_G = 0.057$	0.199
$\mu_{sil} = 2.05$[d]		$\mu_{sil} = 30.82$[d]	

[a] Pooled likelihoods for 6 *Drosophila simulans* and 12 *D. yakuba* strains[22] (771bp; pooling means that within-species log likelihoods are summed).

[b] Likelihoods for 14 *E. coli* strains (1407bp; GenBank[16,17]).

[c] Base frequencies at 4-fold degenerate regular silent sites.

[d] Locus-wide silent mutation rate scaled by N_e (see text).

strains that are significantly different from the others, then model parameters will be estimated incorrectly. For example, this study began with 16 strains of *E. coli* for *gnd,* of which two (labeled r4 and r16) are about as distant from the other *E. coli* strains as they are from *Salmonella*. The remaining *E. coli* strains had an estimated phylogeny that had a more regular appearance. The two aberrant *E. coli* strains were excluded from the analysis.

Estimating Mutation *and* Selection Rates

We now consider a model that will allow us to estimate both the mutation rate μ and the relative selection rate γ for mutants, both scaled by the haploid effective population size N_e. This model is sensitive to saturation and homoplasy, but should give reliable results if the estimate for μ_{sil} (the parameter μ for regular silent sites) is comparable to or greater than the more accurate estimate of μ_{sil} based on the Dirichlet density (1) of the previous section (which, however, assumes selective neutrality at silent sites). This model will be applied both for bases at regular silent sites and for amino acids at codon positions.

Consider a flux of mutations (at the rate μ per generation) in the population. Each mutation changes a base in one individual. Most of the resulting new mutant alleles quickly go extinct by drift, but some survive to have appreciable base frequencies in the population. We ignore subsequent mutations at that site. Since we are assuming that all bases (or amino acids) that are not the ancestral base are selectively equivalent, we can ignore mutation between mutant bases at the same site.

Now view the *population frequencies at those sites* for the surviving mutant bases as a *point process* of frequencies on [0,1]. Under diffusion approximation conditions, this will be a Poisson point process with expected density[10]

$$2\mu\frac{1-e^{-2\gamma(1-p)}}{1-e^{-2\gamma}}\frac{dp}{p(1-p)} \qquad \text{(Selection)}$$

$$2\mu\frac{dp}{p} \qquad \text{(No selection; i.e. } \gamma = 0\text{)}$$

for $0 < p < 1$. In the first case, mutant bases have a relative selective advantage of γ scaled by N_e. Note that these densities are not integrable at $p = 0$. This corresponds to the fact that the population contains a large number of rare mutants at any one time.

Under these assumptions, the counts of the numbers of silent sites that have r sequences with bases different from the ancestral base in a sample of n aligned DNA sequences are independent Poisson with means

$$2\mu \int_0^1 \frac{1 - e^{-2\gamma(1-p)}}{1 - e^{-2\gamma}} \binom{n}{r} p^r (1-p)^{n-r} \frac{dp}{p(1-p)}, \qquad 1 \leq r \leq n-1$$

$$\frac{2\mu}{r} \qquad \text{(No selection; i.e. } \gamma = 0\text{)}$$

We use the counts for polymorphic silent sites within a species to estimate parameters μ_{sil} and γ_{sil} for silent sites, and the counts for amino-acid polymorphic codon positions to estimate parameters μ_{rep} and γ_{rep} for replacement amino acids. We again use maximum likelihood estimates, but lump together both the counts and the probabilities for r and $n - r$ since we may not know the ancestral base or amino acid (Table 3).

The scaled selection rate $\gamma_{rep} = -3.66$ for *E. coli* in Table 3 corresponds to a selection rate of $s = -\gamma_{rep}/N_e$ per generation against replacements, where N_e is the effective population size of *E. coli*. We estimate N_e as follows. The value $\mu_{sil} = 30.82$ in Table 2 corresponds to $N_{sil} \times \mu N_e$, where μ is the mutation rate per site per generation and N_{sil} is the number of amino-acid monomorphic codon positions with twofold or fourfold degenerate regular silent sites. The 14 strains of *E. coli* in Table 2 have $N_{sil} = 367$, and the estimate $\mu = 5 \times 10^{-10}$ per generation[24] implies $N_e = 1.7 \times 10^8$.

This estimate of N_e leads to $s = -\gamma_{rep}/N_e = 2.2 \times 10^{-8}$ per generation against replacements. Thus the average magnitude of selection per generation that acts against observed amino acid substitutions is quite small. One way in which such a small selection coefficient could be realized is if a substitution is selectively neutral in most environments, but disadvantageous in some rarely-encountered environments.[8]

The estimates $\mu_{rep} = 12.51$ (for 469 codons, or $2 * 469 = 938$ first and second codon position sites) but $\mu_{sil} = 30.82$ (for 367 codons) in Table 3 suggests that

Table 3. Estimates of the mutation rate μ and the selection rate γ within one species

14 *E. coli* strains, *gnd* locus (1407p):	
$\mu_{sil} = 30.82$	(From Table 2)
$\mu_{sil} = 33.57 \pm 5.50^{a}$	
$\gamma_{sil} = -1.34 \pm 0.83^{**ac}$	
$\mu_{rep} = 12.51 \pm 4.47^{b}$	
$\gamma_{rep} = -3.66 \pm 2.24^{***bc}$	
$\gamma_{sil} \neq \gamma_{rep}$	($P = 0.029$*)

* $P < 0.05$ **$P < 0.01$ ***$P < 0.001$

[a] Estimated from base distributions at polymorphic regular silent sites.

[b] Estimated from amino acid distributions at amino-acid polymorphic codon positions.

[c] The ranges ± are 95% normal-theory confidence intervals, while P-values are for likelihood ratio tests against $\gamma = 0$.

only about one sixth of amino acid positions in *E. coli* are susceptible to a weakly-selected replacement. This is roughly consistent with the estimate $s = 1.6 \times 10^{-7}$ against replacement amino acids, which is about seven times as large as the value $s = 2.2 \times 10^{-8}$ obtained above, for a similar *E. coli* model in which *all* codon positions in *gnd* were assumed susceptible to a weakly-selected amino-acid replacement.[15]

The closeness of the two estimates of μ_{sil} in Tables 2 and 3 suggests that saturation or homoplasy do not have a significant effect in the estimates of Table 3. The fitted values for counts at regular silent sites for the data in Table 3 are not significantly different from the observed counts ($P = 0.71$, 5 degrees of freedom, 143 polymorphisms). The fitted values for counts for replacement amino acids resembled the observed counts, but had too many empty cells to carry out a chi-square goodness-of-fit test. A more detailed description of the method of estimation in Table 3 will appear elsewhere.

Fixed Differences Between Two Species

Fixed differences between samples are sites (or amino acids) that are fixed within each sample, but fixed at different bases. Given aligned sequence data from two related species, the fixed differences provide additional data that can be used to estimate mutation and selection rates.[10]

The first step in using this additional data is to estimate the time since the two species diverged. Griffiths (1979)[25,26] derived a time-dependent version of the steady-state Dirichlet density in equation (1) for population frequencies. Griffiths' formula gives the joint probability density of two sets of population base frequencies, one for each species, in terms of two sets of $\alpha_T = 2N_e\mu_T, \ldots$ parameters (for the two species), and a parameter t_{div}, which is the scaled number of generations since the two species diverged.

We fix the $\alpha_T, \ldots$ parameters in both species to the pooled estimates of $\alpha_T, \ldots$ from regular silent sites (as in Table 2), and find the maximum likelihood estimate of t_{div} using the joint configurations at regular silent sites for the two samples of DNA sequences aligned together.[10] The estimate of t_{div} for the two *Drosophila* species in Table 2 is

$$t_{\text{div}} = 5.81 \pm 2.52 \qquad \text{(ADH)}$$

In the mutational flux model of the last section, the scaled rate of fixation of mutant bases at the population level is

$$\begin{cases} \mu\dfrac{2\gamma}{1 - e^{-2\gamma}} & \text{(Selection)} \\ \mu & \text{(Noselection; i.e. } \gamma = 0) \end{cases} \qquad (3)$$

If we ignore fixations at the same site in both species, the number of fixed differences will be a Poisson random variable whose mean is $2t_{\text{div}}$ times the relevant expression in equation (3). Technically, the rates in equation (3) are the rates of new mutations that will eventually become fixed. Thus these estimates of fixed differences will overestimate the observed numbers by the number of ancestral polymorphisms that become fixed at different bases, and will underestimate the observed numbers of fixed differences by the number of mutations that will eventually become fixed but remain polymorphic in the present.[10] We ignore both types of errors.

We now obtain maximum likelihood estimates for μ and γ jointly by using polymorphic sites within each species, the estimates for t_{div} above, and the probabilities of fixation by time t_{div} from equation (3) (see Table 4).

The fitted values matched the observed polymorphism counts fairly well, but had a tendency to overestimate counts in one species and underestimate them in the other species.

The mutational flux model assumes either than you combine the observed and theoretical counts for r and $n - r$ differences from the consensus (assuming n DNA sequences), or else that you be able to identify the ancestral base. We followed the first strategy in Table 3 for estimates within each of the species, but attempted to estimate the ancestral bases using the joint configuration data for the two species in Table 4. If both species had the same consensus base at a site, then that base was assumed to be ancestral. The likelihood was summed for the two bases at fixed differences (which eliminates the need to estimate the ancestral base between those two bases), but this was not practical for all polymorphic sites with different consensus bases in the two species. The ancestral bases at those sites were assigned randomly in Table 4 for those sites, and appeared to give similar estimates for different random assignments.

Other strategies would be to combine the counts over r and $n - r$ within each species, as in Table 3, or perhaps to change the assignment with the sign of γ so that the most common base at that site is the one that is selected. The random-assignment strategy may not be the best one. The optimum strategy on this point will be the object of further research.

Table 4. Estimages of μ and γ using two species

6 *D. simulans* and 12 *D. yakuba* strains (771bp; ADH[22]):	
$\mu_{sil} = 2.05$	(From Table 2)
$\mu_{sil} = 2.48 \pm 0.78^{a}$	
$\gamma_{sil} = -0.53 \pm 0.70^{ac}$	$(P = 0.16)$
$\mu_{rep} = 0.068 \pm 0.050^{b}$	
$\gamma_{rep} = 3.58 \pm 9.51^{bc}$	$(P = 0.12)$
$\gamma_{sil} \neq \gamma_{rep}$	$(P = 0.051)$

** $P < 0.01$ ***$P < 0.001$

[a,b,c] See footnotes to Table 3.

*Table 5. A "McDonald-Kreitman[22]" table for ADH (*D. simulans *and* D. yakuba[a]*)*

	fixed at diff. bases	poly. in either spp.
Silent (regular only)	17	21
Replacement (sites)	6	0
	$P = 0.022$[b]	

[a] 6 strains of *D. simulans* and 12 strains of *D. yakuba* (771bp).[22]

[b] Two-sided Fisher exact test.

A McDonald-Kreitman Table

McDonald and Kreitman[22] introduced a 2 × 2 contingency table for selective differences between silent sites and replacement sites in aligned samples between two species. They include the number of sites that are fixed differences and the number of sites that are polymorphic within either species (or both) as the two columns. In the first row we take the numbers of regular silent sites (that are fixed differences or polymorphic) and the similar number of sites within amino acid polymorphic codon positions in the second row. If there is no selective difference between silent and replacement sites, then the 2 × 2 table should be nonsignificant (Table 5).

The contingency table in Table 5 is significant using regular silent sites ($P = 0.022$), while the two-species analysis in Table 4 just misses significance ($P = 0.051$). However, the methods of analysis are quite different, and it would be quite possible for all of the estimates in Table 4 to be significant while the McDonald-Kreitman table above is not significant.

Acknowledgments

This work was partially supported by National Science Foundation Grant DMS-9108262 and National Institutes of Health research grant GM-44889.

References

1. R. C. Lewontin, and J. L. Hubby. (1966). *Genetics* 54, 595–609.
2. H. Harris. (1966). *Proc. Royal Soc. London Ser. B* 164, 298–310.
3. R. C. Lewontin. (1991). *Genetics* 128, 657–662.
4. M. Kimura. (1968). *Nature* 217, 624–626.
5. M. Kimura. (1983). *The Neutral Theory of Molecular Evolution*. Cambridge University Press.
6. C. Wills. (1973). *Amer. Naturalist* 107, 23–34.

7. R. C. Lewontin. (1974). *The Genetic Basis of Evolutionary Change*. Columbia University Press.
8. D. L. Hartl. (1989). *Genetics* 122, 1–6.
9. D. L. Hartl and S. A. Sawyer. (1991). *J. Evol. Biol.* 4, 519–532.
10. S. A. Sawyer and D. L. Hartl. (1992). *Genetics* 132, 1161–1176.
11. W.-H. Li and D. Graur. (1991). *Fundamentals of molecular evolution*. Sinauer Associates.
12. D. L. Hartl and A. Clark. (1989). *Principles of population genetics* (2nd Ed) Sinauer Associates.
13. W.-H, Li, C.-I. Wu and C.-C. Luo. (1985). *Mol. Biol. Evol.* 2, 150–174.
14. K. Wolfe, P. Sharp and W.-H. Li. (1989). *Nature* 337, 283–285.
15. S. A. Sawyer, D. Dykhuizen and D. L. Hartl. (1987). *Proc. Nat. Acad. Sci. USA* 84, 6225–6228.
16. D. E. Dykhuizen and L. Green. (1991). *J. Bacteriol.* 173, 7257–7268.
17. M. Bisercic, J. Y. Feutrier and P. R. Reeves. (1991). *J. Bacteriol.* 173, 3894–3900.
18. Y.-X. Fu and W.-H. Li. (1993). *Genetics* 134, 1261–1270.
19. S. Wright. (1949). pp 365–389 in *Genetics, Paleontology, and Evolution,* edited by G. Jepson, G. Simpson, and E. Mayr. Princeton Univ. Press.
20. J. F. C. Kingman. (1980). *Mathematics of Genetic Diversity* CBMS-NSF Regional Conf. Ser. Appl. Math **34**.
21. G. Watterson. (1977). *Genetics* 85, 789–814.
22. J. H. McDonald and M. Kreitman. (1991). *Nature* 351, 652–654.
23. If you maximize the product of the individual likelihoods (instead of the unknown true joint likelihood), but assume the central limit theorem for the individual log likelihoods, then the maximum likelihood estimator has the correct asymptotic behavior.
24. H. Ochman and A. C. Wilson. (1987). in *Escherichia coli* and *Salmonella typhimurium: Cellular and Molecular Biology* eds J. L. Ingraham, K. B. Low, B. Magasanik, F. C. Neidhardt, M. Schaechter and H. E. Umbarger. (American Society of Microbiology Pubs.).
25. R. C. Griffiths. (1979). *Adv. Appl. Probab.* 11, 310–325.
26. S. Tavaré. (1984). *Theor. Popul. Biol.* 26, 119–164.

8

Detecting Natural Selection by Comparing Geographic Variation in Protein and DNA Polymorphisms

John H. McDonald

Abstract. *Comparing the amount of geographic variation in allele frequencies for protein and DNA polymorphisms is a powerful method for detecting the effects of selection. Some statistical artifacts must be kept in mind, however. Simulations indicate that estimators of Wright's* F_{ST} *are much better measures of geographic variation than are genetic distance measures; that pooling alleles so that all polymorphisms are treated as two-allele polymorphisms is sometimes necessary to avoid statistical artifacts; and that for a given total sample size, two or three population samples can be just as efficient at detecting selection as a larger number of smaller samples.*

Introduction

One of the most powerful methods for detecting the effects of natural selection is comparing geographic variation in allele frequency for different polymorphisms. Drift and migration should cause all neutral polymorphisms to have the same expected amount of geographic variation. Of course there will be a large range around this expectation, but if the appropriate statistical tests, using the appropriate measures of geographic variation, reveal that a polymorphism exhibits either significantly less or significantly more geographic variation than a class of presumably neutral polymorphisms, it is evidence for the effects of natural selection. Balancing selection may be detected if it maintains uniform allele frequencies at one polymorphism in a species in which isolated populations exhibit differentiation at other, neutral polymorphisms due to drift. Differentiating selection may be detected if it causes different allele frequencies at one polymorphism in a species in which migration maintains uniform allele frequencies at other, neutral polymorphisms. While it is possible to use this geographic approach on protein data alone under some circumstances,[6,7,15] it is only really practical if there is a class of polymorphisms that are thought to be affected only by neutral processes. Since the vast majority of DNA polymorphisms cause no change in protein sequence, they are an obvious choice.

Two recent studies illustrate the usefulness of comparing geographic variation of protein and DNA polymorphisms. Karl and Avise[5] surveyed four nuclear DNA polymorphisms in the American oyster, *Crassostrea virginica,* on the Atlantic and Gulf coasts of the United States. When compared with previous data on allozyme polymorphisms,[2] a striking pattern emerged: all fourteen allozyme polymorphisms had less variation in allele frequencies between the Atlantic and Gulf coasts than all four DNA polymorphisms (Figure 1). The DNA polymorphisms were restriction sites in essentially random pieces of single-copy DNA, so it is difficult to imagine differentiating selection affecting all four. The logical conclusion, then, is that balancing selection must be maintaining similar allele frequencies at all of the protein polymorphisms surveyed.

Berry and Kreitman[1] examined the alcohol dehydrogenase (*Adh*) locus in *Drosophila melanogaster* along the East Coast of the United States, where there is a well-known cline in allozyme allele frequency. Using restriction enzymes with four-base recognition sites, they identified 23 DNA polymorphisms with overall average frequencies greater than 0.01, along with the allozyme polymorphism. Two polymorphisms had strikingly higher geographic variation, largely due to latitudinal clines in frequency: the F/S allozyme polymorphism and an insertion/deletion variant called ∇1 (Figure 2). This is consistent with differentiating selection on the F/S and ∇1 polymorphisms causing geographic differentiation in the face of extensive gene flow.

The results indicating balancing selection at 14 polymorphic allozymes in oysters,[5] and differentiating selection at *Adh* and ∇1 in *Drosophila melanogaster,*[1] illustrate the tremendous potential for this kind of geographic comparison. It is therefore important to make sure that statistical artifacts will not cause a misleading apparent difference in geographic variation between what are actually just different kinds of neutral polymorphisms. In addition, some experimental designs may be much more efficient at detecting the effects of any selection that is present. Here I consider some of the statistical questions raised by this kind of geographic study.

It is first necessary to choose an appropriate statistical measure of geographic variation. Distance measures, such as Cavalli-Sforza and Edwards' arc distance,[3] Rogers' distance,[12] and Nei's standard genetic distance,[8] are sometimes used to measure between-population variation, but simulations indicate that they are strongly affected by the initial number of alleles at a polymorphism (Figure 3). Clearly, this is not desirable when comparing the amount of geographic variation at different polymorphisms. In addition, distance measures are limited to two samples and cannot be used to measure variation among several populations at once. Wright's F_{ST}[17] is a parameter that can be estimated for any number of population samples, and in simulations it is much less affected by the initial number and frequencies of alleles.

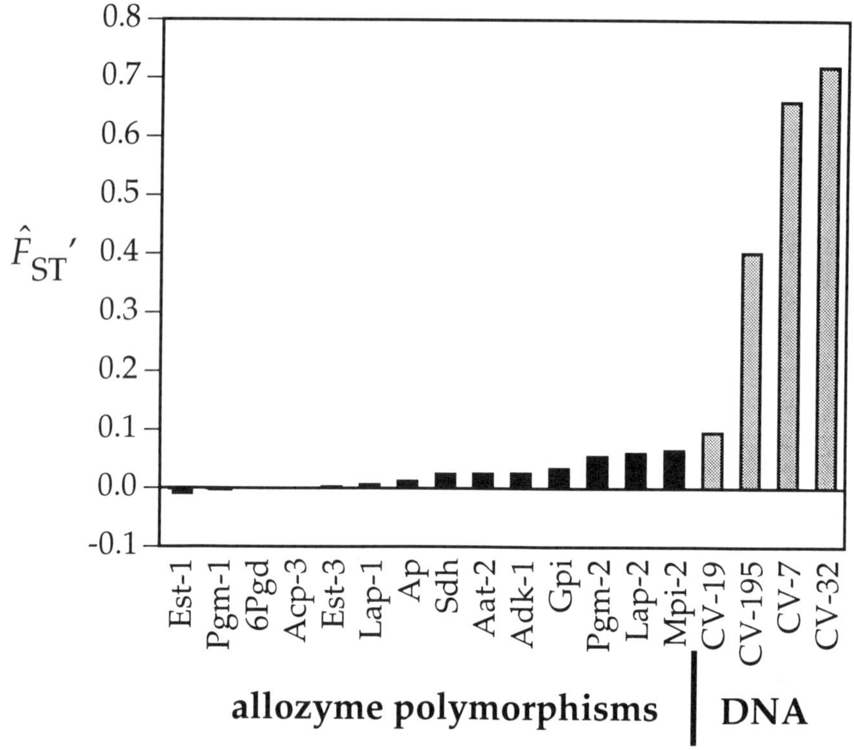

Figure 1. Geographic variation in the American oyster, *Crassostrea virginica,* between the Atlantic and Gulf coasts of the United States. Allele frequencies for allozyme polymorphisms[2] were taken for the nine locations geographically closest to the nine locations at which nuclear DNA polymorphisms were sampled.[5] Data from five Atlantic Coast samples were pooled, as were data from four Gulf Coast samples. All but the most common allele for allozyme polymorphisms were pooled, to make two-allele polymorphisms comparable to the two-allele DNA polymorphisms. $\hat{F}_{ST}'$ was calculated according to Nei.[9] $\hat{F}_{ST}'$ was significantly different ($P<0.001$) between the allozyme and DNA polymorphisms (Mann-Whitney *U*-test; Sokal and Rohlf 1981, pp. 433–434), indicating that at least one class of polymorphisms is affected by selection.

Because F_{ST} is a population parameter, not a sample statistic, it is necessary to use a statistical estimator of F_{ST}. Protein polymorphisms as assayed using allozyme electrophoresis can easily be determined in thousands of individuals, while DNA polymorphisms are generally determined for smaller numbers, so it is important to use a statistic that removes the effects of different sample sizes at different loci. Treating F_{ST} as a sample statistic, as has sometimes been done in allozyme studies, is therefore inappropriate; doing so conflates both geographic variation and sampling variation, so F_{ST} is increased for small sample sizes

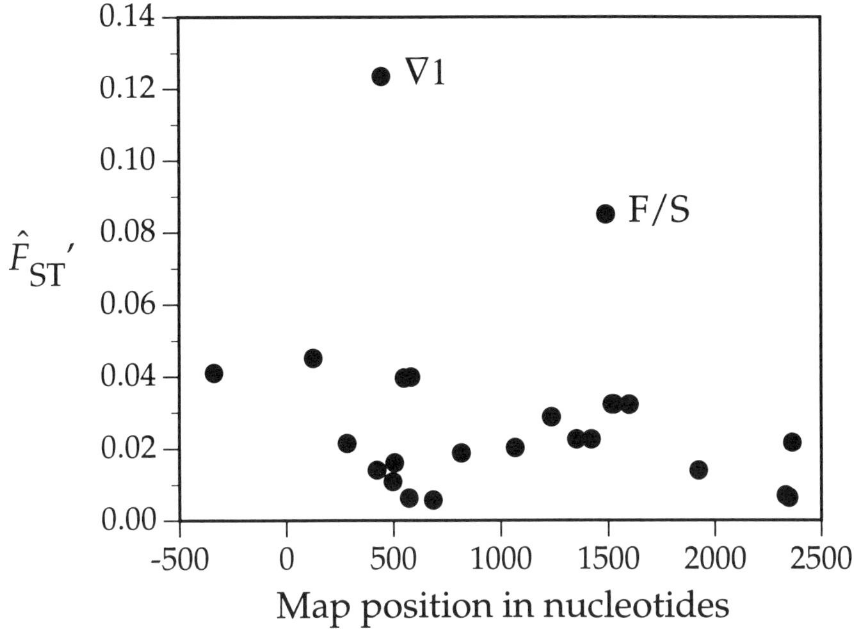

Figure 2. $\hat{F}_{ST}'$ for DNA polymorphisms at the *Adh* locus in *Drosophila melanogaster*, sampled at 15 locations on the East Coast of the United States. Data are from Berry and Kreitman.[1] $\hat{F}_{ST}'$ was calculated for all polymorphisms where the average frequency of the rare allele was greater than 0.01, and only samples with data for all polymorphisms were included. Samples from single locations in different years are kept separate, so the $\hat{F}_{ST}'$ includes some temporal as well as geographic variation. ∇1 is an insertion of 34 basepairs, combined with a deletion of 30 basepairs, and F/S is the nucleotide substitution responsible for the allozyme polymorphism.

(Figure 4). DNA polymorphisms might then appear to exhibit greater geographic variation than protein polymorphisms just because the DNA sample sizes were smaller. There are two commonly used statistical estimators of F_{ST} for two-allele polymorphisms: $\hat{\theta}$[16] and $\hat{F}_{ST}'$.[9] In simulations of two drifting, isolated populations, both with equal population sizes, $\hat{\theta}$ and $\hat{F}_{ST}'$ are virtually identical when equal sample sizes are taken from the two populations, and both estimators are relatively unaffected by wide variation in sample size (Figure 4). When unequal sample sizes are collected from the two populations, however, differences between the methods of accounting for sample size become apparent (Figure 5). For the particular conditions simulated here, a polymorphism with one small and one large population sample would appear to have a greater $\hat{\theta}$, and a smaller $\hat{F}_{ST}'$, than a polymorphism with two equal sample sizes. Simulations of a great variety of situations would be needed to determine how serious this artifact is, but it is clear that neither estimator of F_{ST} does a perfect job of removing the effects of

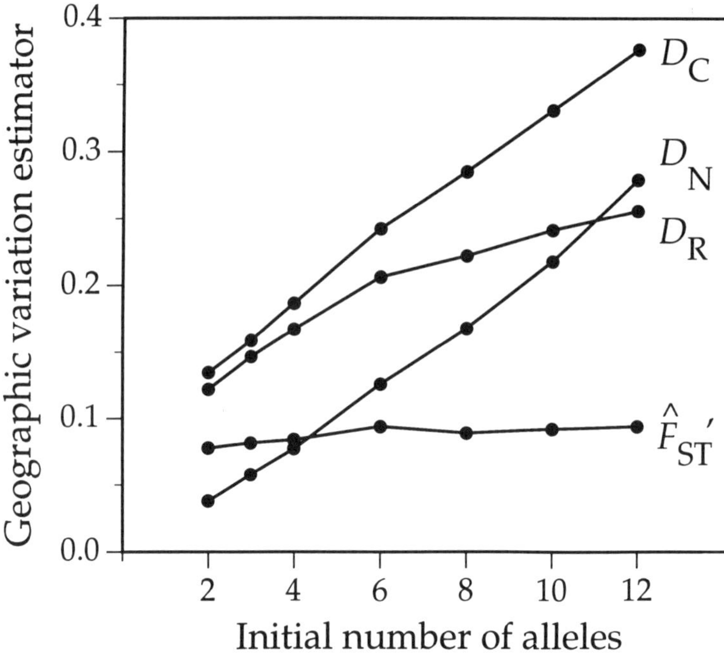

Figure 3. Effect of the number of alleles on estimators of geographic variation. Random drift without migration or mutation was simulated for two populations of 100 haploid individuals each, starting with equal initial allele frequencies in the two populations and drifting for 10 generations. Random sets of initial allele frequencies were drawn from the distributions expected at equilibrium for the given number of alleles and 100 individuals,[4] using the algorithm of Stewart.[14] After the simulated random drift a sample of 100 individuals was drawn at random with replacement, and the following measures of geographic variation were estimated: D_C, Cavalli-Sforza and Edwards' arc distance[3]; D_N, Nei's standard genetic distance[8]; D_R, Rogers' distance[12]; and $\hat{F}_{ST}'$, Nei's estimator of Wright's F_{ST}.[9] Two thousand replicates were run for each number of alleles, and each replicate had a new set of initial allele frequencies. Replicates in which both populations became fixed for the same allele were omitted, and the means of each estimator were then calculated.

different sample sizes in different populations. To avoid this artifact, it will be desirable to use similar sample sizes from each population sampled whenever possible.

It is also important that the estimator of geographic variation be unaffected by the number and proportions of alleles at a polymorphism. Protein polymorphisms often consist of several alleles, many of which are rare, while DNA variation such as presence/absence of restriction sites are two-allele polymorphisms. In other studies with different kinds of data, one might want to compare

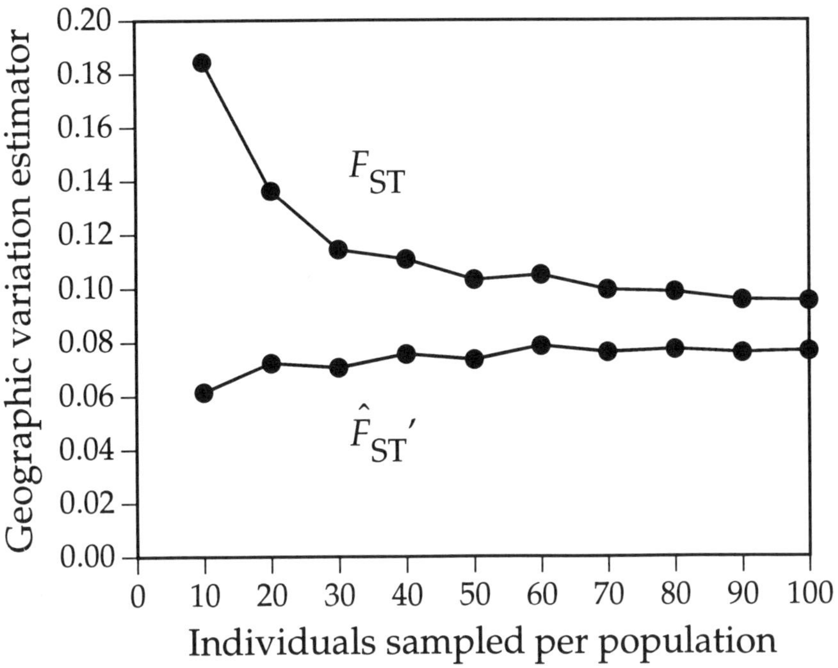

Figure 4. Effect of sample size on estimators of geographic variation. Random drift was simulated as in Figure 3, except that only two-allele polymorphisms were simulated. A sample of the given number of haploid individuals was then drawn at random with replacement, and F_{ST}[17] and $\hat{F}_{ST}'$[9] were determined. $\hat{\theta}_W$[16] was also determined, but it is indistinguishable from $\hat{F}_{ST}'$ at this scale.

a two-allele protein polymorphism with haplotype data based on a number of restriction sites, so that there were many DNA haplotypes at a single locus. Simulations of two randomly drifting, completely isolated populations show that the method of pooling estimates of F_{ST} across alleles at a single locus has an effect on the mean of the estimator. $\hat{F}_{ST}'$[9] and $\hat{\theta}_W$ are weighted averages of the F_{ST} estimator for each allele, where alleles with intermediate frequencies are weighted more heavily. These estimators exhibit smaller means for two-allele polymorphisms than for polymorphisms with greater numbers of alleles (Figure 6). One can take an unweighted average of $\hat{\theta}$ or $\hat{F}_{ST}'$ across alleles; the unweighted average of $\hat{\theta}$ is $\hat{\theta}_U$,[16] and the unweighted average of $\hat{F}_{ST}'$ will here be called $\hat{F}_{ST(U)}$, found by calculating $\hat{F}_{ST}'$[9] for each allele at a locus separately, then taking the average across alleles. These unweighted averages remove much of the bias in the means (Figure 6). The distributions of $\hat{\theta}_U$ or $\hat{F}_{ST(U)}$ for two- vs. many-allele polymorphisms are different, however, so that two-allele polymorphisms will have a greater range of geographic variation than multiple-allele polymorphisms (Figure 7). If a multi-allele protein polymorphism is compared with

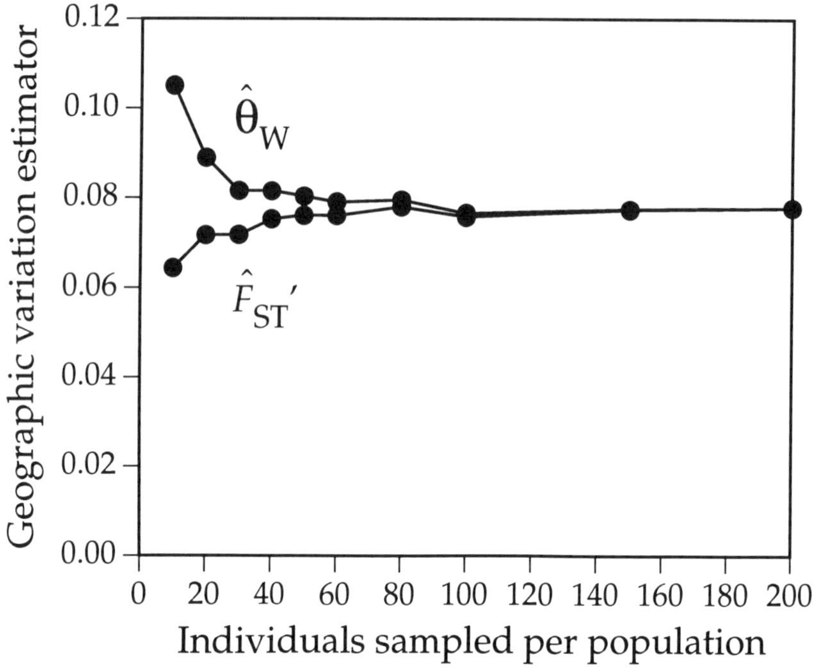

Figure 5. Effect of using two different sample sizes on estimators of geographic variation. Random drift was simulated as for Figure 4. A sample of the given number of haploid individuals was then drawn at random from one population, and a sample of 200 individuals was drawn from the other population. $\hat{\theta}_W$[16] and $\hat{F}_{ST}'$[9] were determined, and the means from two thousand replicates for each sample size are shown.

many two-allele DNA polymorphisms, the smaller variance for the multi-allele polymorphisms would make the test conservative. However, comparing a two-allele protein polymorphism with multi-allelic DNA polymorphisms could be misleading. For the simulated conditions, 22 percent of the two-allele polymorphisms have $\hat{F}_{ST(U)}$ that would appear to be significantly different than the multi-allele polymorphisms (Figure 7). For this kind of comparison, it would be better to treat all polymorphisms as two-allele polymorphisms, by using the frequencies of the overall most common allele and pooling all the others.

There are many other conditions that could be simulated, but at this point it appears that so long as there is not a consistent difference in number of alleles between protein and DNA polymorphisms, measuring geographic variation with $\hat{F}_{ST(U)}$ or $\hat{\theta}_U$ is relatively unbiased. Unless there are large differences in sample size among locations, the difference between $\hat{F}_{ST(U)}$ and $\hat{\theta}_U$ is minuscule; I will use $\hat{F}_{ST(U)}$ here, since it is somewhat simpler computationally.

It is then important to decide what kind of statistical test to use, before collecting the data. If $\hat{F}_{ST(U)}$ was normally distributed, one could compare a single

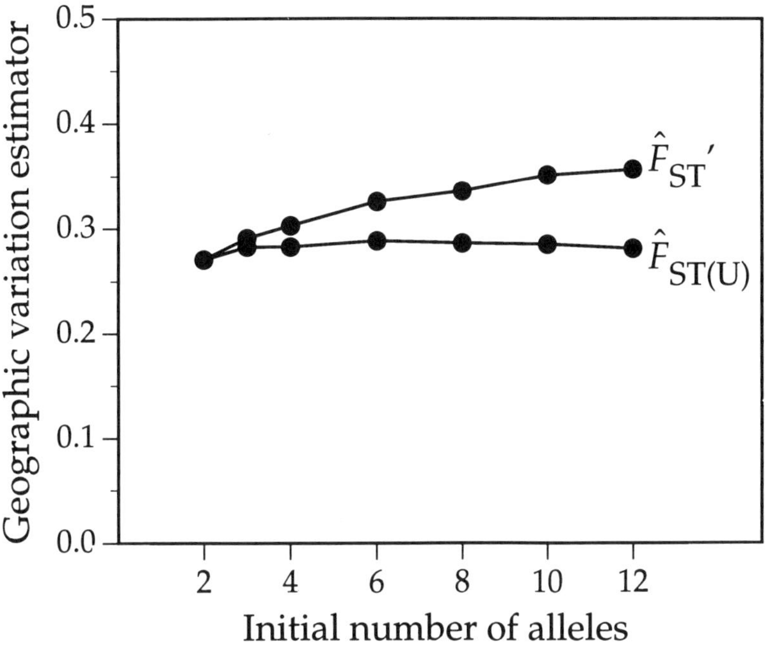

Figure 6. Effect of the number of alleles on estimators of geographic variation. Random drift was simulated as for Figure 3, except that 50 generations of drift were simulated. A sample of 100 individuals was then drawn at random with replacement from each population, and $\hat{F}_{ST}'$[9] and $\hat{F}_{ST(U)}$ (the unweighted average across alleles of $\hat{F}_{ST}'$) were determined. $\hat{\theta}_W$ and $\hat{\theta}_U$[16] are indistinguishable from $\hat{F}_{ST}'$ and $\hat{F}_{ST(U)}$, respectively, at this scale.

protein polymorphism to a small number of DNA polymorphisms using the t-test for a single observation compared with a mean (Sokal and Rohlf 1981, p. 231). Unfortunately, $\hat{F}_{ST(U)}$ is not normally distributed (Figure 7), and the shape of the distribution will depend on the number and size of populations,[10] the amount of migration,[11] and selection on linked polymorphisms. The shape of the distribution would be difficult to estimate from a small number of polymorphisms. It may be possible to develop a parametric test using some kind of bootstrapping approach, but for now a simple non-parametric test seems appropriate: if a protein polymorphism has a smaller $\hat{F}_{ST(U)}$ than 97.5 percent of DNA polymorphisms, it is evidence for balancing selection, and if a protein polymorphism has a greater $\hat{F}_{ST(U)}$ than 97.5 percent of DNA polymorphisms, it is evidence for differentiating selection. To test for selection on a single protein polymorphism, it thus would be necessary to collect data on at least 39 DNA polymorphisms. Obtaining 39 DNA polymorphisms is a daunting task; one could do a one-tailed statistical test and get by with 19 DNA polymorphisms, but this

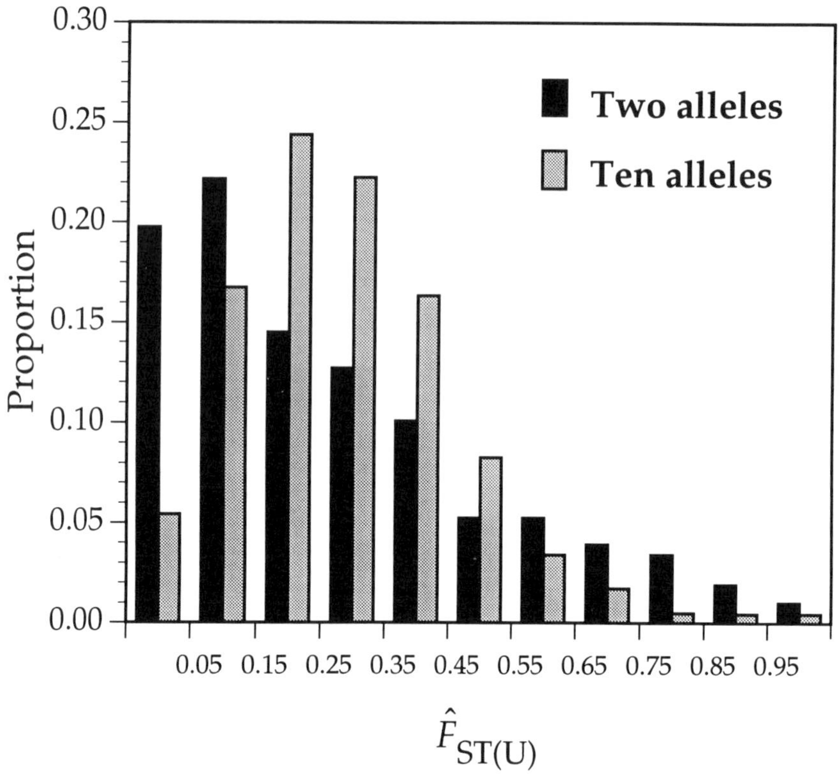

Figure 7. Effect of number of alleles on the distribution of $\hat{F}_{ST(U)}$. Two populations of 100 haploid individuals each, drifting for 50 generations, were simulated as in Figure 6. The proportion of the 2000 replicates within each range of $\hat{F}_{ST(U)}$ is shown for two kinds of loci: polymorphisms initially with two alleles, and polymorphisms initially with ten alleles. (The polymorphisms that initially had ten alleles had an average of 3.6 alleles after the 50 generations of drift.) Although the mean $\hat{F}_{ST(U)}$ is about the same for two- and ten-allele polymorphisms (0.270 and 0.285, respectively), the two-allele polymorphisms have a much broader distribution. Of the two-allele polymorphisms, 12.6 percent have a smaller $\hat{F}_{ST(U)}$ than 97.5 percent of the ten-allele polymorphisms, and 9.4 percent of the two-allele polymorphisms have a greater $\hat{F}_{ST(U)}$ than 97.5 percent of the ten-allele polymorphisms.

would only be appropriate if an *a priori* prediction had been made of the direction of difference in $\hat{F}_{ST(U)}$ that would result from selection.

We next would like to get an idea of the number of population samples to take, and how large to make those samples. The obvious answer is "as many samples, with as many individuals, as possible," but time and money may limit the number of individuals it is feasible to assay for DNA polymorphisms. It is then necessary to determine whether it is better to take a small number of population samples, each with a large number of individuals, or to take a larger number of population samples, each with a smaller number of individuals. Often this will depend on what other questions about population structure one wishes to address, but some simulations of ideal situations may be helpful.

First consider the situation where several populations, all of the same size, have been completely isolated from each other for the same number of generations. One is looking for the pattern where a protein polymorphism has a smaller $\hat{F}_{ST(U)}$ than 97.5 percent of neutral DNA polymorphisms, due to balancing selection. By doing simulations to find the number of generations of drift required for 97.5 percent of neutral polymorphisms to have a greater $\hat{F}_{ST(U)}$ than the median polymorphism with uniform allele frequencies, it appears that sampling three populations is considerably more powerful than sampling two populations, for the same total sample size (Figure 8). Spreading the same number of individuals across more than three populations does not increase the ability to detect selection, under these conditions. If some of the populations are not as diverged as others, due to larger population size, greater migration or shorter time of isolation, samples from the less diverged populations will contribute relatively little to the $\hat{F}_{ST(U)}$, and in that case sampling the most diverged population and one other is about as powerful as sampling several populations, for the same total sample size. Of course, this would require knowing *a priori* which population is likely to be most diverged due to drift.

The other possible pattern indicative of selection is a protein polymorphism that exhibits greater geographic variation than 97.5 percent of neutral DNA polymorphisms, due to diversifying selection. The amount of divergence required for a protein polymorphism to have a greater $\hat{F}_{ST(U)}$ than 97.5 percent of neutral polymorphisms that have not diverged at all is only a function of the total sample size, not the number of populations it is divided into (Figure 9). One therefore might just as well sample only two populations, if one is confident that they have enough migration to keep neutral polymorphisms uniform in frequency, yet are different enough in environment that adaptive protein polymorphisms might have diverged.

The discussion so far has indicated that the optimal experimental design is often to sample only two populations, carefully chosen so that they will have diverged due to drift either a lot, or not at all. In a species that has already been surveyed for geographic variation in enzyme polymorphisms, there are many possible pairwise comparisons of samples. Even in a species in which drift has caused extensive differentiation, some pairs of locations will be virtually identical in allele frequency for a particular allozyme polymorphism, just by chance. If one picked the two locations that were most similar in frequency for a particular enzyme polymorphism and then surveyed DNA polymorphisms from those two locations, it would not be surprising if the DNA polymorphisms all had greater geographic variation, even if the protein polymorphism was also neutral. Similarly, in a species where allele frequencies are really kept uniform by migration,

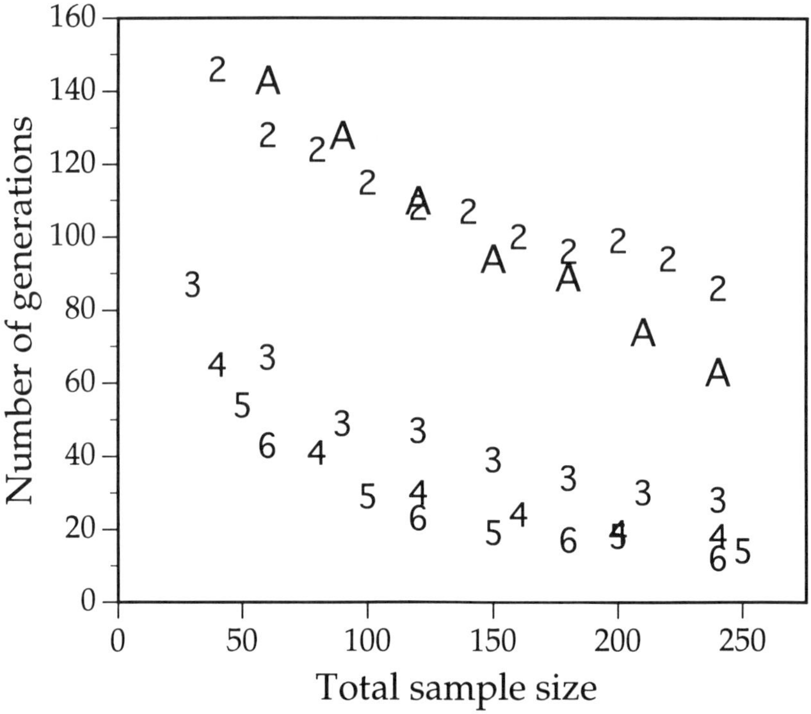

Figure 8. Number of generations of drift required for a protein locus with allele frequencies kept uniform by balancing selection to have significantly lower $\hat{F}_{ST(U)}$ than neutral polymorphisms. Random drift was simulated as in Figure 4 for the number of populations indicated by the symbol, and $\hat{F}_{ST(U)}$ was calculated every ten generations. The median $\hat{F}_{ST(U)}$ for a protein locus with uniform allele frequencies, sampled with 200 individuals per population, was also calculated. By interpolation, the number of generations of drift required for 97.5 percent of neutral loci to have a greater $\hat{F}_{ST(U)}$ than the median protein was determined. "Total sample size" is the number of haploid individuals sampled, totaled across all populations. The symbols indicate the number of populations simulated. "A" represents three populations, two drifting for the given number of generations plus one population drifting for half as long. The similarity between it and the results for two populations hints that the advantage of sampling three or more populations may disappear if not all populations are the same size and have been isolated for the same length of time.

there will be some variation in observed frequencies due to sampling error. If one deliberately picked the pair of locations that differed the most for a particular enzyme polymorphism, then examined DNA polymorphisms from those two locations, one might see that the DNA polymorphisms had less geographic variation and erroneously conclude that this was evidence for selection.

Clearly, either approach would be misleading. Ideally, it would be best to sample protein and DNA polymorphisms in a species in which neither was

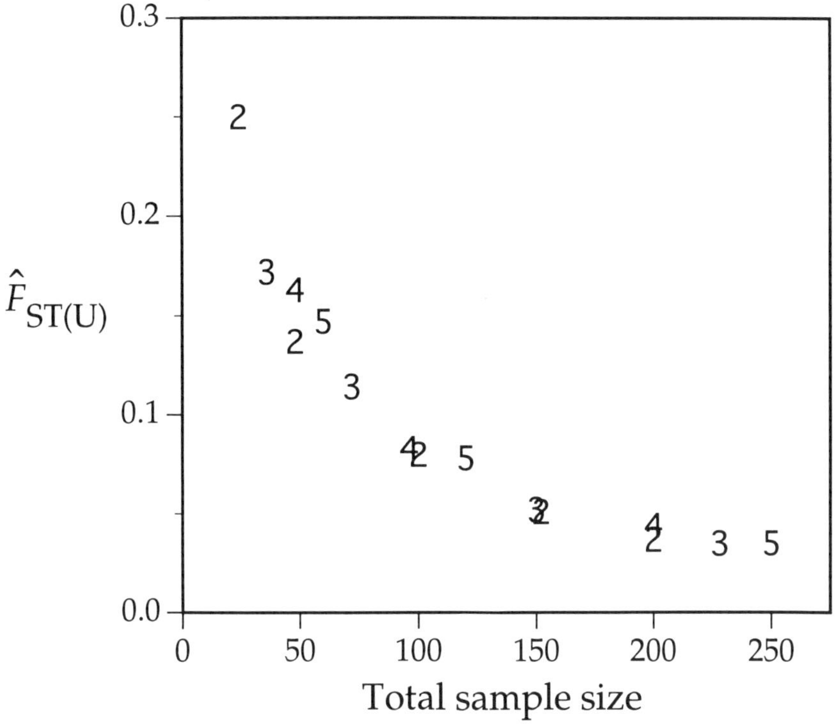

Figure 9. $\hat{F}_{ST(U)}$ of a differentially selected protein locus that would be greater than 97.5 percent of neutral loci with allele frequencies kept uniform by migration. A single population with a two-allele polymorphism was simulated with an initial allele frequency for the less-common allele drawn at random between 0.05 and 0.50. The total number of haploid individuals indicated was sampled from this population, they were divided into the number of samples indicated by the symbol, and $\hat{F}_{ST(U)}$ was calculated. One thousand replicates were conducted for each number of samples, each with a new initial allele frequency. The replicates in which a single allele was fixed in all samples were discarded, and then the $\hat{F}_{ST(U)}$ greater than that of 97.5 percent of the neutral polymorphisms was determined.

previously studied, or at least to sample polymorphisms at locations which hadn't been previously sampled. Since allozyme electrophoresis is a dying art, it will be tempting to collect DNA data from locations where enzyme polymorphisms were sampled previously. It would then be necessary to either choose the locations without regard to their allozyme allele frequencies, or to sample all of the locations from which allozyme data were previously collected.

References

1. A. Berry, and M. Kreitman. (1993). Molecular analysis of an allozyme cline: alcohol dehydrogenase in *Drosophila melanogaster* on the East Coast of North America. *Genetics* **134,** 869–893.

2. N. E. Buroker. (1983). Population genetics of the American oyster *Crassostrea virginica* along the Atlantic coast and the Gulf of Mexico. *Marine Biology* **75,** 99–112.
3. L. L. Cavalli-Sforza, and A. W. F. Edwards. (1967). Phylogenetic analysis: models and estimation procedures. *Evolution* **21,** 550–570.
4. W. J. Ewens. (1972). The sampling theory of selectively neutral alleles. *Theoretical Population Biology* **3,** 87–112.
5. S. A. Karl, and J. C. Avise. (1992). Balancing selection at allozyme loci in oysters: implications from nuclear RFLPs. *Science* **256,** 100–102.
6. R. C. Lewontin, and J. Krakauer. (1975). Testing the heterogeneity of *F* values. *Genetics* **80,** 397–398.
7. J. H. McDonald. (1991). Contrasting amounts of geographical variation as evidence for direct selection: the *Mpi* and *Pgm* loci in eight crustacean species. *Heredity* **67,** 215–219.
8. M. Nei. (1972). Genetic distance between populations. *American Naturalist* **106,** 283–292.
9. M. Nei. (1986). Definition and estimation of fixation indices. *Evolution* **40,** 643–645.
10. M. Nei, and A. Chakravarti. (1977). Drift variances of F_{ST} and G_{ST} statistics obtained from a finite number of isolated populations. *Theoretical Population Biology* **11,** 307–325.
11. M. Nei, A. Chakravarti and Y. Tateno. (1977). Mean and variance of F_{ST} in a finite number of incompletely isolated populations. *Theoretical Population Biology* **11,** 291–306.
12. J. S. Rogers. (1972). Measures of genetic similarity and genetic distance. pp. 145–154 in *Studies in genetics: VII,* M. R. Wheeler, ed. University of Texas, Austin.
13. R. R. Sokal, and F. J. Rohlf. (1981). *Biometry*. W. H. Freeman, San Francisco. 859 pp.
14. F. M. Stewart. (1977). Computer algorithm for obtaining a random set of allele frequencies for a locus in an equilibrium population. *Genetics* **86,** 482–483.
15. S. Tsakas and C. B. Krimbas. (1976). Testing the heterogeneity of *F* values: a suggestion and a correction. *Genetics* **84,** 399–401.
16. B. S. Weir, and C. C. Cockerham. (1984). Estimating *F*-statistics for the analysis of population structure. *Evolution* **38,** 1358–1370.
17. S. Wright. (1951). The genetical structure of populations. *Annals of Eugenics* **15,** 323–354.

9

A Neutrality Test for Continuous Characters Based on Levels of Intraspecific Variation and Interspecific Divergence

Andrew G. Clark

Abstract. *One of the most powerful approaches for inferring the action of evolutionary forces is the comparison of intraspecific nucleotide polymorphism to the divergence between species in DNA sequences. The infinite sites model of neutral molecular evolution provides a well-described null hypothesis for expected levels of polymorphism and divergence, and the Hudson-Kreitman-Aguadé test is one formal statistical test for departure from this null hypothesis. In this paper, the predictions of the neutral model for continuous characters are examined, as are some data on intraspecific variation and interspecific divergence of metabolic characters. These traits are being examined because of the likely simplicity of the molecular basis for their variation. Predictions of within-population variance and between-population divergence for quantitative characters have been derived by several investigators, and here these results are brought together in the context of testing the null hypothesis of neutrality. Data on variation in enzyme activities provide clear departures from the neutral model. Some limitations of the approach are discussed.*

Introduction

Molecular population genetics has been very successful at challenging the view that most change at the molecular level occurs by neutral substitution. Arguably the most important tool in this challenge has been the recognition that the neutral theory makes predictions about both polymorphism within species and the degree of nucleotide sequence divergence between species. Statistical tests which are essentially goodness-of-fit tests to neutrality have been devised on this principle,[8] and several significant departures have been seen. An even simpler test to implement makes the additional inference that under neutrality the levels of polymorphism and divergence ought to be the same for both synonymous and nonsynonymous sites.[15] This test has revealed cases in which there appears to be a large excess of amino acid substitutions that occur as species diverge, an observation that has been interpreted as evidence for positive Darwinian evolution.

The goal of this paper is to explore the idea of comparing intraspecific variation and interspecific divergence for quantitative traits. This idea is not new. In an influential paper, Kluge and Kerfoot[10] showed that morphological characters that exhibit the highest level of intrapopulation variance also show the greatest level of interpopulation divergence. This phenomenon can be tested in several ways statistically, but it appears to be fairly robust.[17] What was missing from these early papers is an attempt to relate the phenomenon to the underlying quantitative genetic basis for the characters, and to thereby give a plausible evolutionary model for the phenomenon. The contribution of this paper is to illustrate a unity between the Kluge-Kerfoot phenomenon and the Hudson-Kreitman-Aguadé test, and to show that this unity arises from the theory of evolutionary quantitative genetics. The theory will require consideration of an underlying genetic basis of the quantitative characters, and will provide inferences about those underlying genes from the changes in the distributions of the phenotypes. In this particular case, the underlying genetic variation will be assumed to be following strictly neutral random genetic drift.

Polymorphism and Divergence of Neutral Continuous Characters

Much of the theory of drift and quantitative characters that we need has been developed by Lande,[11] Lynch and Hill,[14] and by Felsenstein.[6] Assume discrete generations, a population of N diploid individuals, random mating, and no selection. The rate at which mutation inflates the variance in a character each generation is $V_m = \Sigma 2n\mu \mathrm{E}(a^2)$, where n is the number of genes, μ is the per locus mutation rate, and the allelic effect of a new mutation is a. This quantity is also known as the mutational variance. Such a population will come to an equilibrium such that the rate of gain of variance by mutation is matched by the rate of loss of variance due to random genetic drift.

One can derive the equilibrium variance for an additive trait in the manner of Clayton and Robertson.[4] First note that the recursion for the expected variance is

$$E[V_w(t)] = (1 - 1/2N)V_w(t-1) + V_m \quad (1)$$

This recursion makes it clear that variance is lost at a rate of $1-1/2N$, as it must at each neutral locus, and variance is increased by V_m each generation. The general solution for this recursion is

$$E[V_w(t)] = V_w(1 - 1/2N)^t + 2NV_m[1 - (1 - 1/2N)^t] \quad (2)$$

and as $t \rightarrow \infty$, it is clear that the solution is

$$E[V_w] = 2NV_m. \quad (3)$$

Lynch and Hill[14] consider the case of dominance, letting $\bar{k}$ be the average degree of dominance (additive case is $\bar{k} = 0$) and σ_k^2 be the variance in degree of dominance. The variance within a population at the mutation-drift equilibrium is:

$$E[V_w] = 2NV_m[1 + (2/3)\bar{k}^2 + \bar{k} + \sigma_k^2) \quad (4)$$

In the case where there is no dominance, so that allelic effects at each locus are additive, we get equation (3), as expected.

Lynch and Hill also calculate the rate of divergence of the mean for a pair of populations that are totally isolated. In this case, we have to specify the number of generations of isolation (t), because, in the absence of selection or some other stabilizing force, the variance between populations is expected to grow without bound. In the case of additivity,

$$E[V_b(t)] = 4NV_m\{(t/2N) - [1 - \exp(-t/2N)]\}. \quad (5)$$

For small t and large N, this is approximately $2V_m t$ (a result also found by Lande[11]). Note that rate of divergence does not depend on population size, just as for the molecular clock for neutral nucleotide sequence evolution (Figure 1). This is reassuring, since the cause for the inflation of variance between isolated populations is random genetic drift of the underlying genes. Lynch and Hill point out that, on this time scale, effects of dominance, linkage and linkage disequilibrium can be virtually ignored. They justify this claim through extensive numerical simulations on the computer.

Extension to Multiple Characters

Life is seriously complicated by the fact that phenotypic characters do not evolve independent of one another. Whereas we can often get by with simple models of nucleotide evolution which treat substitutions as independent, for phenotypic characters, a model of independence makes little sense. The reason is simply that the underlying genetic variation almost always has pleiotropic effects. In order to test whether the degree of variation within a population and divergence between populations fits a neutral expectation, these genetic correlations must be taken into account. Fortunately, there is a multiple-trait analog to the model derived above. If **M** is the covariance matrix of mutational effects, then the balance between influx of mutations and loss of variation through drift occurs with a genetic covariance matrix within species of:

$$\mathbf{G}_w = 2N\mathbf{M}. \quad (6)$$

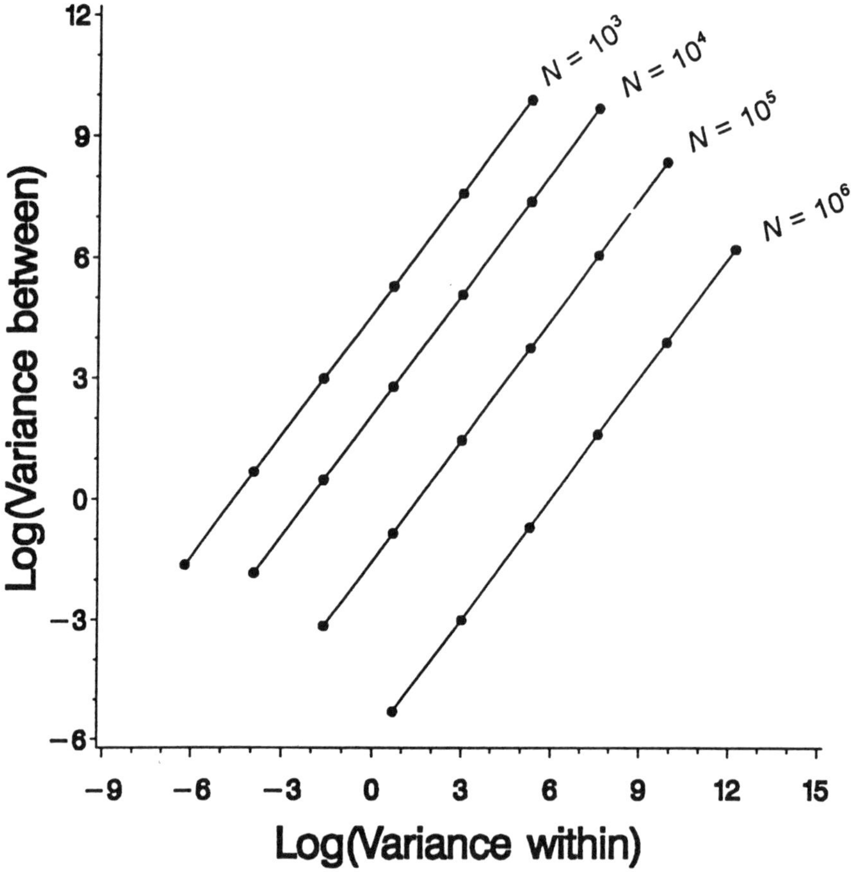

Figure 1. Expected variance within a population and between two populations for a trait that is determined by additive, strictly neutral alleles in a population with population size $N = 10^3 - 10^6$, and mutational variance $V_m = 10^{-6}$ (bottom of each diagonal line) to 10^{-1} (top of each diagonal line). Note that for a given population size, different traits might have different V_ms, but they should fall on a diagonal line, with a correlation between variance within a population and divergence between.

Genetic covariances are increased and decreased with the same underlying dynamics of neutral drift at constituent alleles as is genetic variance. For the same reason, the covariance matrix between diverging species has the dynamic

$$\mathbf{G}_b = 4N\mathbf{M}\,[t/2N - (1 - \exp(-t/2N)]. \tag{7}$$

So the between-species covariance matrix is expected to be proportional to the within species covariance matrix. This provides us with the necessary information we need to construct a test for goodness of fit.

Analog to the Hudson-Kreitman-Aguadé Test

The basic idea of the test is that characters with high intraspecific variation ought to exhibit the greatest divergence between species. In the multivariate context, under pure drift, we expect the divergence between species to be determined by the covariance matrices of the characters within the two species. The equations that determine the covariance matrices have as parameters the mutational covariance matrix **M**, the population size N and the time of divergence t. The Hudson-Kreitman-Aguadé test actually estimates the parameters $\theta = 4N\mu, f$ (the relative population sizes of the two species) and the divergence time t by least squares. These parametric estimates are then used to formulate a chi square statistic for a goodness-of-fit test to the strictly neutral model. This approach is unlikely to be useful in the context of continuous characters, because there are insufficient degrees of freedom. Instead, we will simply test the concordance of the within and between species covariance matrices, or between individual terms of the covariance matrices. The approach is best demonstrated with some real data.

The Data

Clark and Keith[2] devised means of estimating enzyme activities in 96-well microtiter plates, and the precision of the methods was tested with samples of 350 field-caught *D. melanogaster* and 354 field-caught *D. simulans*. The characters that were measured in this study included the live weight of the flies, the amounts of stored lipid (LIP) and glycogen (GLY), and the activities of enzymes of fatty acid synthase (FAS), glucose-6-phosphate dehydrogenase (G6PD), glycogen phosphorylase (GP), α-glycerol phosphate dehydrogenase (GPDH), glycogen synthase (GS), hexokinase (HEX), malic enzyme (ME), 6-phosphogluconate dehydrogenase (PGD), phosphoglucose isomerase (PGI), phosphoglucomutase (PGM), and trehalase (TRE). Other studies had demonstrated the utility of treating enzyme activities as continuous, quantitative traits, and significant variation was found between extracted chromosome lines of *Drosophila* even if they share the same electromorph.[1,13] Many pairs of enzyme activities also show a pattern of strong genetic correlations among extracted lines,[18] reflecting metabolic constraints. The data from Clark and Keith[2] were used in several tests of concordance of within-species variance and between-species divergence.

First, univariate tests were done to ascertain that the 14 traits were approximately normally distributed in both species. We could then estimate the mean and within-species variance for each character. The square of the difference between species means was then plotted against the within-species variance, and the results show a striking correlation (Figure 2, Spearman rank correlation $\rho = 0.874$, $P < 0.0001$). At first sight, this would appear to be consistent with the classical Kluge-Kerfoot effect, but the power in detecting departures from a

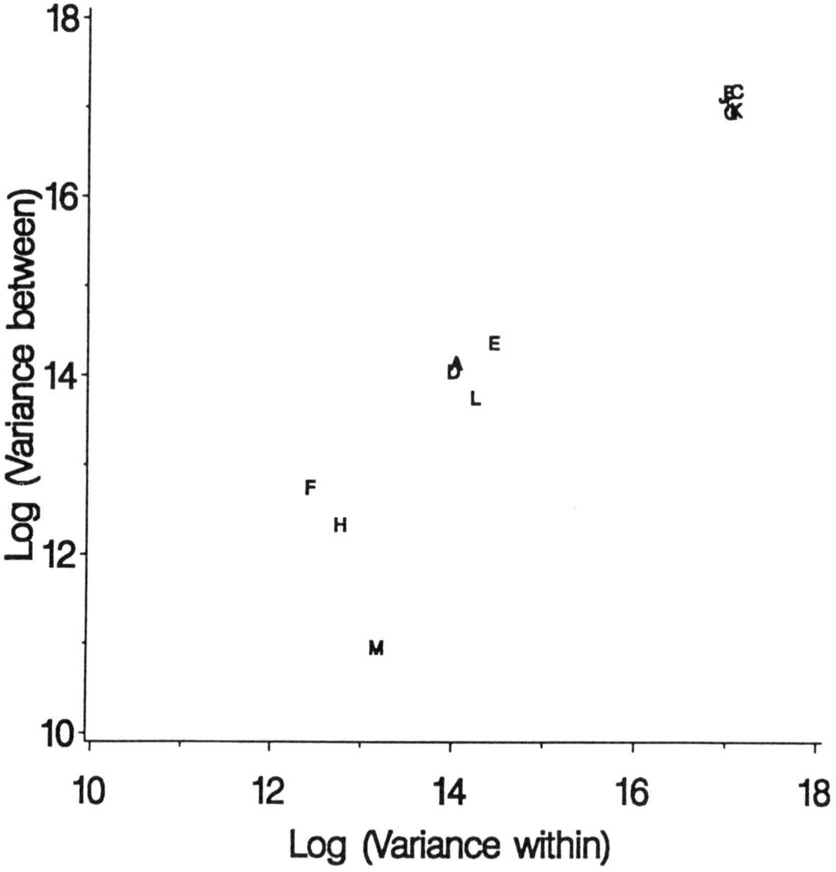

Figure 2. Correlation between intraspecific variance and interspecific divergence for a series of metabolic traits in *D. melanogaster* and *D. simulans*. The 13 characters are A:FAS, B:G6PD, C:GP, D:GPDH, E:GS, F:HEX, G:ME, H:PGD, I:PGI, J:PGM, K:TRE, L:LIP, M:GLY. The Spearman rank correlation is $\rho = 0.874$ ($P < 0.0001$). This rough correspondence between divergence and within-species variance appears to be consistent with neutrality, but the test is lacking in power.

line is quite poor, and the test illustrated in Figure 2 only considers single traits, while much of the interest in the data lies in the consistency of covariances between traits.

The theory developed above suggests that the covariance matrices for within and between species variation should be proportional. There are many tests of equivalence of covariance matrices, and the likelihood ratio test[16] indicates a highly significant rejection of the null hypothesis that the within and between species covariance matrices are proportional. This result implies that the neutral null hypothesis is not adequate to explain the pattern of variance and divergence that is seen. Rather than belabor the details of this test, note that it is not really very informative because any one element in the matrix could be responsible for the departure. We would like to know which particular characters are responsible for the departure.

An *ad hoc* bootstrapping approach was used to test the concordance of within and between species variance for each term in the covariance matrices. One thousand bootstrap samples were constructed, and from each, a within-species covariance matrix and a between-species covariance matrix was constructed. Each bootstrap sample consisted of drawing 500 pairs of individuals with replacement from the data set. For pairs that were of the same species, the difference between the two individuals was calculated for each trait, and sums of squares and crossproducts of these differences were tallied for the within-species covariance matrix **C.** If the pair of individuals were from the two different species, then, as for the within-species matrix, the difference between the two individuals was calculated for each trait, and sums of squares and crossproducts of these differences were tallied for the covariance matrix of species divergence, **D.** Each bootstrap sample of 500 pairs yields two covariance matrices, and we calculate the ratio of divergence: within-species covariance, D_{ij}/C_{ij}, for each term. From the set of bootstrap samples, we calculate a mean and standard error of these ratios for each element, and these provide a simple test of the null hypothesis that the terms in the divergence covariance matrix are proportional to the terms in the within-species covariance matrix. The results show that many terms, particularly the variance terms, deviate from the null hypothesis, and we can reject a model of simple neutrality as the cause of the patterns of divergence and variation (Figure 3). For example, the figure shows that weight exhibits an excess in divergence compared to the within-species variance (*simulans* were heavier), and that the WT × TRE covariance was significantly diverged. Note that the test can only detect excess divergence, and that strong selection conserving a particular trait or covariance is not distinguished from neutrality.

Phylogenetic Comparative Approach

The basic idea of scaling expected rates of divergence to the level of intrapopulation variance makes the test described above sensitive to many perturbing factors besides natural selection. The ubiquity of pleiotropic effects forces us to consider correlated effects. The phylogenetic comparative approach is another method that tests a neutral null hypothesis for pairs of traits by asking whether the changes in the traits occurs on a phylogeny in an independent fashion.[5,7,9] Tests of this sort are different from the above in that they do not require information on intraspecific variation, but they do require data on multiple species for which a phylogeny is known. An underlying model behind one of the common methods

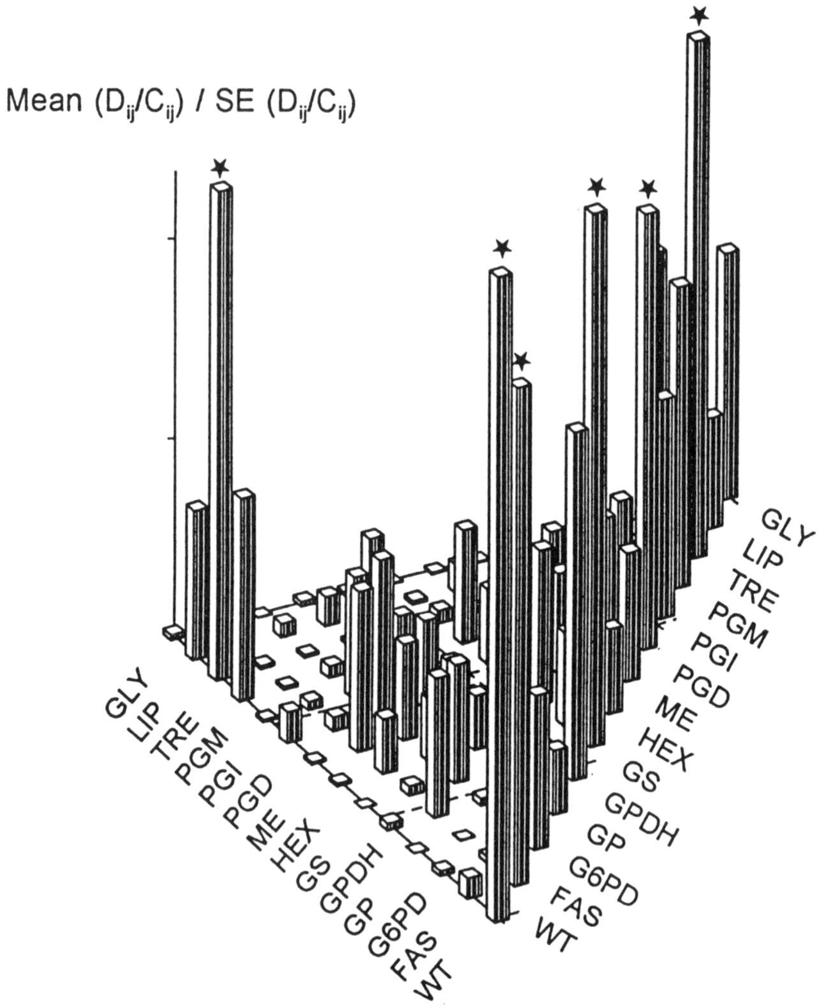

Figure 3. Bootstrap test for concordance of individual terms in the covariance matrix of metabolic characters in *D. melanogaster* and *D. simulans*. **D** is the covariance matrix of differences between pairs of individuals drawn from two different species, and **C** is the covariance matrix of differences between pairs of individuals from the same species. The x and y axes are simply the character names, and the z axis is the ratio of mean to standard deviation of the distribution of bootstrap samples for D_{ij}/C_{ij}. Asterisks indicate a significant departure from neutrality.

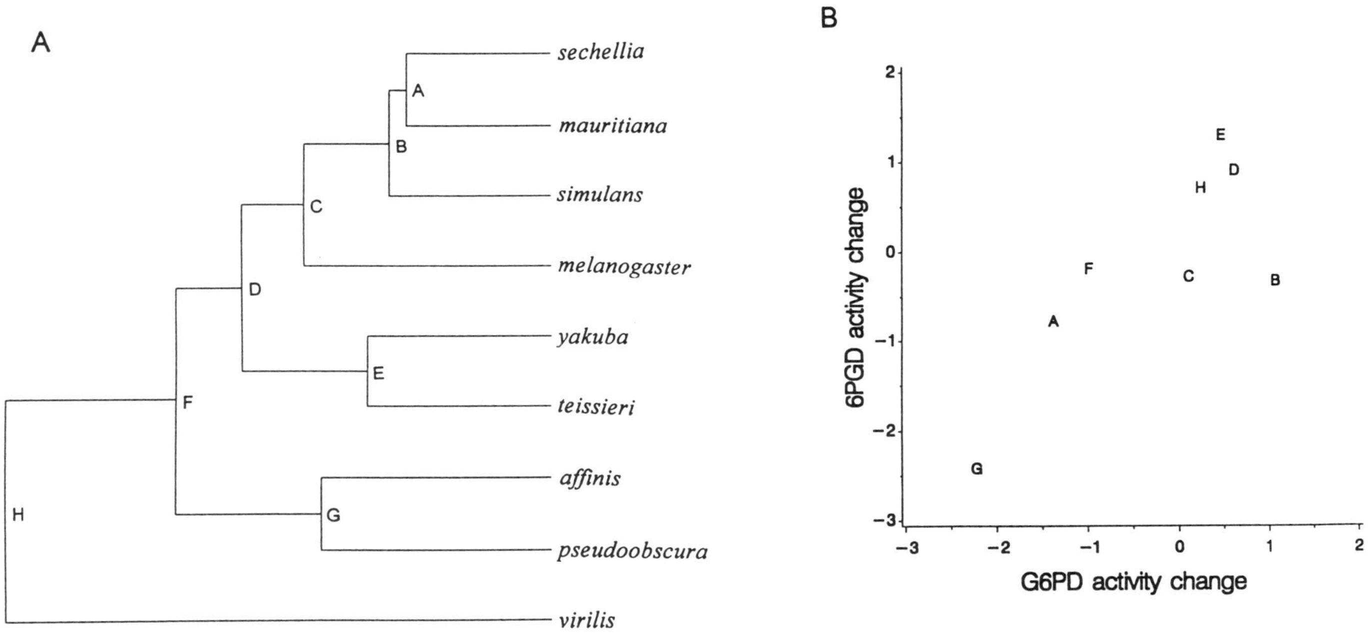

Figure 4. (A) UPGMA tree obtained by pooling a variety of molecular data, including mtDNA, rRNA, DNA-DNA hybridization, and sequences of several nuclear genes (see Clark and Wang[3]). The node labels are used in the method of independent comparisons, discussed in the text. (B) Changes in activity of G6PD and 6PGD on the two sides of each node whose labels appear in Figure 6. The points exhibit a significant positive trend, as tested by a Spearman rank correlation ($\rho = 0.83$, $P < 0.01$). This implies that the neutral null hypothesis, that changes in G6PD and 6PGD occur independently, can be rejected. Instead, it appears that the coregulation of G6PD and 6PGD is conserved on the phylogeny.

for phylogenetic comparisons assumes that traits change by Brownian motion as gene frequencies wander by genetic drift.[5,6] Under this model, if two traits are not functionally related, then they should appear to be uncorrelated across independent lineages. As described above, covariance over species between pairs of traits ought to reflect the mutational covariance if the traits are neutral.[11] The comparative method may provide means for ascertaining whether covariance between traits is due to ancestry or function.[7]

We applied this method to the metabolic characters examined in the analysis of Figures 2 and 3 on nine species of *Drosophila,* and the results appear in Figure 4 (see also Clark and Wang[3]). Interspecific comparisons were made using the inferred phylogenetic relations between the species. For each node in the phylogenetic tree of Figure 4A, the mean of each character in the species above and below each node was calculated. For each pair of traits, we then plot the difference in traits means between the species above vs. below each labeled node. Figure 4B illustrates the result for the pair of traits G6PD and 6PGD activity. We are particularly interested in these traits because they are the enzymes leading in the pentose phosphate shunt, and other data demonstrate a genetic correlation between these traits within *D. melanogaster*. The significant phylogenetic correlation would not occur by drift alone, but represents an evolutionary constraint on the coordinated expression of these two enzymes.

In summary, the idea of polymorphism and divergence have been applied to continuous characters, and large discrepancies in within- and between-species variation can lead to a rejection of a strictly neutral model. The understanding of adaptation at both the molecular and the phenotypic levels can benefit from analysis that make use of phylogenetic relations to compare variation within a species to divergence between species.

Acknowledgments

I thank Lei Wang for technical assistance. This work was supported by grant BSR 9007436 from the National Science Foundation.

References

1. A. G. Clark, and L. E. Keith. (1988). Variation among extracted lines of *Drosophila melanogaster* in triacylglycerol and carbohydrate storage. *Genetics* **119,** 595.
2. A. G. Clark and L. E. Keith. (1989). Rapid enzyme kinetic assays of individual *Drosophila* and comparisons of field-caught *D. melanogaster* and *D. simulans*. *Biochem. Genet.* **27,** 263–277.
3. A. G. Clark and L. Wang. (1993). Comparative evolutionary analysis of metabolism in nine *Drosophila* species. *Evolution, in press.*

4. G. A. Clayton and A. Robertson. (1955). Mutation and quantitative variation. *Am. Nat.* **89,** 151–158.
5. J. Felsenstein. (1985). Phylogenies and the comparative method. *Am. Nat.* **125,** 1–15.
6. J. Felsenstein. (1988). Phylogenies and quantitative characters. *Ann. Rev. Ecol. Syst.* **19,** 445–471.
7. P. H. Harvey and M. D. Pagel. (1991). *The Comparative Method in Evolutionary Biology*. Oxford University Press, Oxford.
8. R. R. Hudson, M. Kreitman and M. Aguadé. (1987). A test of neutral molecular evolution based on nucleotide data. *Genetics* **116,** 153–159.
9. R. B. Huey and A. F. Bennett. (1987). Phylogenetic studies of coadaptation: Preferred temperatures versus optimal performance temperatures of lizards. *Evolution* **41,** 1098–1115.
10. A. G. Kluge and W. C. Kerfoot. (1973). The predictability and regularity of character divergence. *Am. Nat.* **107,** 426–442.
11. R. Lande. (1976). Natural selection and random genetic drift in phenotypic evolution. *Evolution* **30,** 314–334.
12. R. Lande. (1980). The genetic covariance between characters maintained by pleiotropic mutations. *Genetics* **94,** 203–215.
13. C. C. Laurie-Ahlberg, G. Maroni, G. C. Bewley, J. C. Lucchesi, and B. S. Weir. (1980). Quantitative genetic variation of enzyme activities in natural populations of *Drosophila melanogaster*. *Proc. Natl. Acad. Sci. USA* **77,** 1073–1077.
14. M. Lynch and W. G. Hill. (1986). Phenotypic evolution by neutral mutation. Evolution **40,** 915–935.
15. J. H. McDonald and M. Kreitman. (1991). Adaptive protein evolution at the *Adh* locus in *Drosophila*. *Nature* **351,** 652–654.
16. D. F. Morrison. (1976). *Multivariate Statistical Methods,* Second edition, McGraw-Hill, New York, pp. 252–253.
17. R. Sokal. (1976). The Kluge-Kerfoot phenomenon reexamined. *Am. Nat.* **110,** 1077–1091.
18. A. N. Wilton, C. C. Laurie-Ahlberg, T. H. Emigh, and J. W. Curtsinger. (1982). Naturally occurring enzyme activity variation in *Drosophila melanogaster*. II. Relationship among enzymes. *Genetics* **102,** 207–221.

10

Estimation of Population Parameters and Detection of Natural Selection from DNA Sequences

Wen-Hsiung Li and Yun-Xin Fu

Abstract. *A great difficulty in population genetics is our inability to measure any of the basic parameters that are involved in the theory of population genetics. For example, we do not know the long-term effective population size (N_e) of any species and the mutation rate μ per sequence per generation for any gene. Fortunately, the situation is improvingly as a result of the introduction of various molecular techniques into population genetics. In particular, the feasibility of obtaining a fairly large number of DNA sequences from a population has enabled us to obtain better estimates of certain parameters such as* $\theta = 4N_e\mu$.

Another challenging problem to population geneticists is how to detect the presence of natural selection in a DNA region, i.e., to test the hypothesis that a DNA region is subject to no selective constraints. Again, this problem has become simpler because of our ability to obtain a large sample of sequences from a population.

The aim of this article is to review the mathematical theory that has been developed in recent years for dealing with the above two problems. The theory is based largely on the coalescent theory. Hudson[8] *gave an excellent review of the theory.*

Gene Genealogy and the Coalescent Theory

Suppose that a random sample of n sequences of a DNA region is taken from a diploid random mating population of effective size N_e. If the DNA region is completely linked so that no recombination occurs between sequences, then the n sequences are connected by a single phylogenetic tree, i.e., a genealogy (Figure 1). In other words, the n sequences can be traced back first to $n-1$ ancestral sequences, next to $n-2$ ancestral sequences and so on until reaching a single common ancestral sequence. Let t_m be the time duration (the number of generations) required for the coalescence from m sequences to $m-1$ sequences. Under the neutral Wright-Fisher model, that is, the population under study evolves

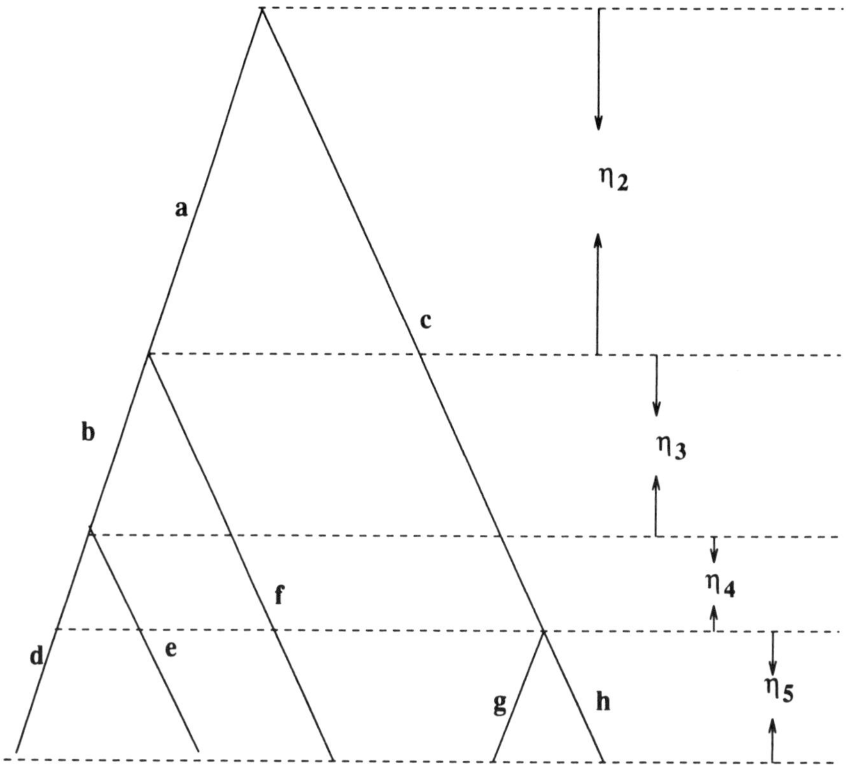

Figure 1. An example of genealogy of five genes. η_k ($k = 2, \ldots, 5$) is the number of mutations during the period that there were k ancestral genes in the genealogy.

according to the Wright-Fisher model, the DNA region is selectively neutral and there is no recombination and no population subdivision. Kingman[9], Hudson[7] and Tajima[12] showed that for $m \geq 2$, t_m has the exponential distribution

$$g(t_m) = \frac{m(m-1)}{M} \exp\left(-\frac{m(m-1)}{M} t_m\right) \tag{1}$$

where $M = 4N_e$ and different coalsecent times, t_i and t_j $(i \neq j)$, are independent of each other. If the coalescent times are scaled so that one unit corresponds to M generations, then the scaled coalescent time $t'_m(= t_m/M)$ follows the exponential distribution

$$g(t'_m) = m(m-1)e^{-m(m-1)t'_m} \tag{2}$$

The mean and variance of the coalescent time t_m are

$$E(t_m) = \frac{M}{m(m-1)} \tag{3}$$

$$Var(t_m) = \left[\frac{M}{m(m-1)}\right]^2 \tag{4}$$

and the mean and variance of the scaled coalescent time t'_m are

$$E(t'_m) = \frac{1}{m(m-1)} \tag{5}$$

$$Var(t'_m) = \left[\frac{1}{m(m-1)}\right]^2 \tag{6}$$

The genealogy of n sequences has 2(n−1) branches. A branch is said to be external if it directly connects to an external node, otherwise it is said to be internal. In Figure 1, branches d,e,f,g and h are external whereas branches a, b and c are internal. Let J_n, I_n and L_n be, respectively, the total time length of all branches, the total time length of internal branches and the total time length of external branches. Fu and Li[4] showed that

$$E(L_n) = M$$

which is independent of the sample size n. Thus, regardless of the number of sequences sampled, the expected total time length of the external branches is always $M = 4N_e$ generations. The expectations of I_n and J_n are

$$E(I_n) = (a_n - 1)M \tag{7}$$

$$E(J_n) = E(I_n) + E(L_n) = a_n M \tag{8}$$

where

$$a_n = 1 + \frac{1}{2} + \ldots + \frac{1}{n-1} \tag{9}$$

Let us now consider mutations. Assume that the rate of mutation per sequence per generation is μ and that the number of mutations in a sequence in a time period follows the Poisson distribution. Let η_e and η_i be the total number of mutations in the external and internal branches, respectively, and let $\eta = \eta_i + \eta_e$ be the total number of mutations that occurred in the entire genealogy of n sequences. Fu and Li[4] showed that

$$E(\eta_e) = \theta \tag{10}$$

$$Var(\eta_e) = \theta + c_n\theta^2 \tag{11}$$

$$E(\eta_i) = (a_n - 1)\theta \tag{12}$$

$$Var(\eta_i) = (a_n - 1)\theta + \left(b_n - \frac{2a_n}{n-1} + c_n\right)\theta^2 \tag{13}$$

$$E(\eta) = a_n\theta \tag{14}$$

$$Var(\eta) = a_n\theta + b_n\theta \tag{15}$$

where

$$b_n = 1 + \frac{1}{2^2} + \ldots + \frac{1}{(\mathrm{n}-1)^2} \tag{16}$$

$$c_n = \begin{cases} 1, & n = 2 \\ 2\dfrac{na_n - 2(n-1)}{(n-1)(n-2)} & n > 2 \end{cases} \tag{17}$$

Under the infinite-site model, which assumes that every mutation occurs at a new site of the sequences, the number of segregating (polymorphic) sites (K) among the sequences in the sample is equal to the total number of mutations η. Therefore, the mean and variances of η should be the same as those of K and indeed, formula (14) and (15) are identical to those of Watterson[14].

The purpose of studying the above quantities will become clear when we consider the estimation of θ and detection of natural selection in a DNA region. Another quantity of interest is the average number of nucleotide differences between two sequences (Π_n) in a random sample of n sequences. Tajima[12] showed that

$$E(\Pi_n) = \theta \tag{18}$$

$$Var(\Pi_n) = \frac{n+1}{3(n-1)}\theta + \frac{2(n^2+n+3)}{9n(n-1)}\theta^2 \tag{19}$$

Also, it is interesting to study the relationships among these quantities. For example, Figure 2 shows that η_e and η_i become almost independent (i.e., have a low correlation coefficient) when the sample size n becomes larger than 10 whereas the correlation between η and Π_n is strong even when the sample size is quite large.

Potential for Improving the Estimation of θ

As is well known, the parameter θ plays a prominent role in the stochastic theory of population genetics. So there has been much interest in the technical aspects

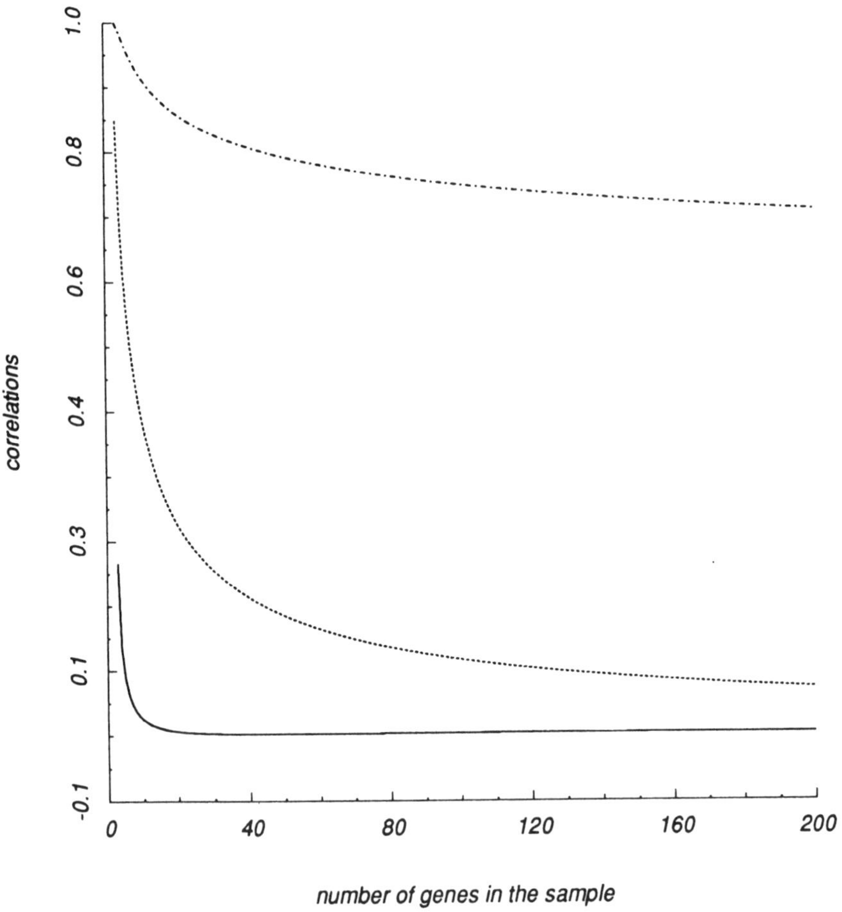

Figure 2. Correlations. The solid, dotted and the dash-dotted curves are the correlations between η_e and η_i, between η_e and Π_n and between η and Π_n, respectively

of estimating θ. In the past two estimators of θ have been commonly used: Watterson's estimator[14] and Tajima's[12] estimator. As can be seen from Eq. (14)

$$\hat{\theta}_W = \eta/a \tag{20}$$

should be an unbiased estimator of θ. As noted above under the infinite-site model, $K = \eta$ and so the estimator defined by Eq. (20) is equivalent to Watterson's estimator. From Eq. (18), one obtains Tajima's estimator of θ as

$$\hat{\theta}_T = \Pi_n \tag{21}$$

It is known that Watterson's estimator is more efficient than Tajima's estimator because it has a smaller variance (e.g. Felsenstein[2]).

The above two estimators actually make little use of the genealogical relationships among the sequences in the sample. Strobeck[11] developed a theory that takes into account the phylogenetic relationships between the alleles in a sample, however, he considered only the cases of two and three alleles. More recently, Felsenstein[2] considered any number of sequences. Under the assumptions that the sequences under study are infinitely long and that the scaled coalescent times t_m's can be inferred without error, he showed that the improvement can be so great that the efficiency of Watterson's estimator approaches zero as the sample size becomes very large. However, Felsentein's assumptions were unrealistic and it was unclear how much improvement can be achieved in practice. Fu and Li[5] addressed this question.

Let us consider the hypothetical genealogy in Figure 1. The tree is divided into segments by the horizontal lines through each of the branching nodes (or tips). Let η_j be the total number of mutations that occurred between the $(j-1)$-th node and the j-th node, $j = 2, \ldots, n$. Suppose that in an idealized situation that we can observe or infer without error the η_j values. Then using the coalescent theory and the maximum likelihood method, one can show that the optimal estimator $\tilde{\theta}$ is the solution of the equation

$$\sum_{m=2}^{n} \frac{\eta_m + 1}{\theta + m - 1} = \frac{\eta}{\theta} \tag{22}$$

and the variance of this estimator is

$$V_{min} = \frac{\theta}{\sum_{m=2}^{n} \frac{1}{\theta + m - 1}} \tag{23}$$

(Fu and Li[5]).

We now compare $\tilde{\theta}$ with Watterson's estimator $\hat{\theta}_W$ which can be obtained from (15). The former is based on $\eta_2, \ldots, \eta_n$ whereas the latter is based on only η, the sum of $\eta_2, \ldots, \eta_n$. Obviously, $\tilde{\theta}$ utilizes more information than $\hat{\theta}_W$ and should be more efficient than $\hat{\theta}_W$. Indeed, Fu and Li[5] showed that

$$Var(\tilde{\theta}) \leq Var(\hat{\theta}_W) \tag{24}$$

Since in practice, the η_j's cannot be estimated without error or in other words, we have less information than knowing $\eta_2, \ldots, \eta_n$, we cannot have a better estimator than $\tilde{\theta}$, i.e., a smaller variance than V_{min}. Thus, V_{min} is a lower bound of variances for all possible unbiased estimators of θ.

The ratio $Var(\tilde{\theta})/Var(\hat{\theta}_W)$ determines the efficiency of $\hat{\theta}_W$ relative to $\tilde{\theta}$ and thus the potential for improving the estimation of θ over Watterson's estimator $\hat{\theta}_W$. Fu and Li[5] showed that

$$\lim_{n\to\infty} \frac{Var(\tilde{\theta})}{Var(\hat{\theta}_W)} = 1 \tag{25}$$

This means that for extremely large sample size n, $\hat{\theta}_W$ becomes as good as $\tilde{\theta}$ or, in other words, $\hat{\theta}_W$ is asymptotically optimal. However, the rate of approaching to optimality is very slow. Indeed, Figure 3 shows that the relative efficiency of $\hat{\theta}_W$ to $\tilde{\theta}$ is considerably smaller than 1 even when the sample size is as large as 500. Thus, there is a large potential for finding a better estimator than $\hat{\theta}_W$.

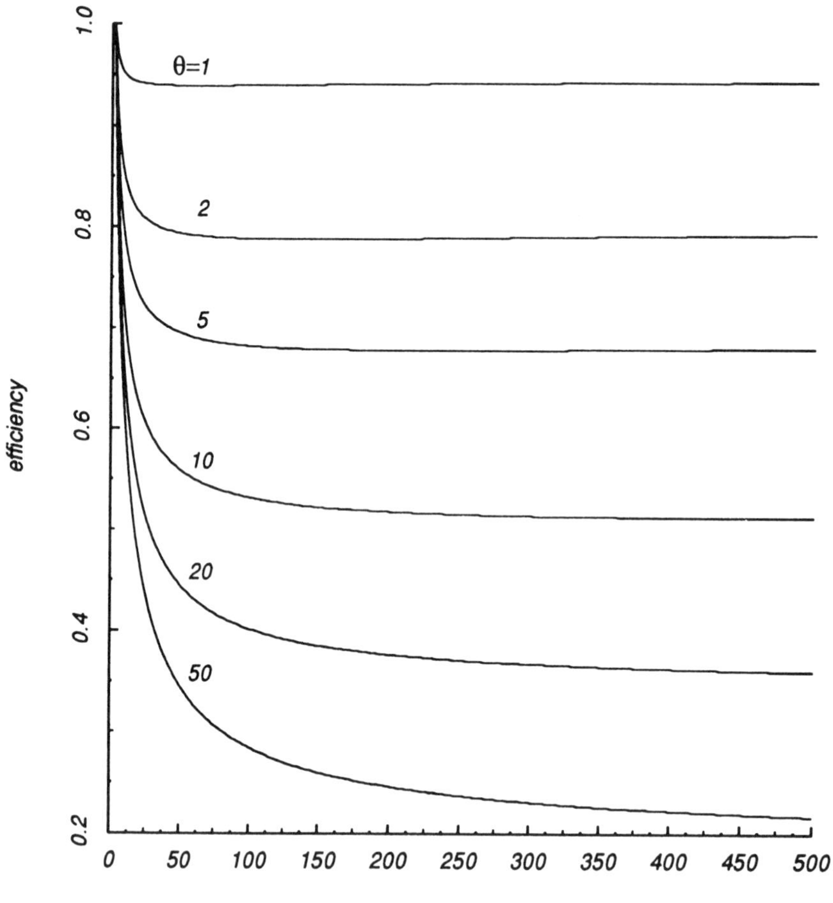

Figure 3. Efficiency of Watterson's estimate $\hat{\theta}_W$ of θ relative to $\tilde{\theta}$. The six curves from top to bottom correspond respectively to θ = 1, 2, 5, 10,20 and 50. The efficiency is calculated by $V_{min}/Var(\hat{\theta}_W)$.

As discussed below, a better estimator can indeed be obtained by inferring a phylogenetic tree from the sequences in the sample.

New Estimator of θ

As demostrated in the previous section, one can obtain an estimator of θ with the minimum variance among all possible unbiased estimators (or nearly unbiased estimators) if the values $\eta_2, \ldots, \eta_n$ can be inferred without error. In reality, the best one can hope is to reconstruct the genealogy of a sample without error. Therefore, a practical approach is to develop a best estimator from information in the genealogy inferred from a sample and such an estimator is presented below.

Let l_i be the scaled time length of branch i (i.e. one unit corresponds to 4N generations) and m_i be the number of mutations on branch i. Assuming that for each i, m_i follows Poisson distribution with parameter θl_i conditional on l_i, we then have for $i \neq j$

$$E(m_i) = \alpha_i\theta$$
$$Var(m_i) = \alpha_i\theta + \beta_{ii}\theta^2$$
$$Cov(m_i,m_j) = \beta_{ij}\theta^2$$

where

$$\alpha_i = E(l_i) \tag{26}$$
$$\beta_{ij} = Cov(l_i,l_j) \tag{27}$$

Because once the genealogy is given α_i and β_{ij} are constants, we can see that the expectation of m_i is a linear function of θ and the variance of m_i and the covariance between m_i and m_j are quadratic functions of θ. Define a vector $\mathbf{m} = (m_1, \ldots, m_{2(n-1)})^T$. Then the mean and variance of $\mathbf{m}$ can be written in the following form:

$$E(\mathbf{m}) = \theta\boldsymbol{\alpha} \tag{28}$$
$$Var(\mathbf{m}) = \theta\boldsymbol{\gamma} + \theta^2\boldsymbol{\beta} \tag{29}$$

where $\boldsymbol{\alpha}$ is a vector defined by $\boldsymbol{\alpha} = (\alpha_1, \ldots, \alpha_{2(n-1)})^T$ and $\boldsymbol{\gamma}$ and $\boldsymbol{\beta}$ are both $2(n-1) \times (n-1)$ matrices defined by

$$\boldsymbol{\gamma} = diag(\alpha_1, \ldots, \alpha_{2(n-1)})$$
$$\boldsymbol{\beta} = \{\beta_{ij} : i,j = 1, \ldots, 2(n-1)\}$$

We can take advantage of the structure of mean and variance of **m** and derive a minimum variance estimator of θ from $m_1, \ldots, m_2(n-1)$. Before we present the theory of estimation, let us look at one example on how α_i's and β_{ij}'s can be computed.

Consider the genealogy of three sequences (Figure 4). The scaled time lengths of the four branches are given by

$$l_1 = l_2 = t_3'$$
$$l_3 = t_3' + t_2'$$
$$l_4 = t_2'$$

It follows from (5) and (6) that $\alpha_1 = E(t_3') = 1/6$, $\beta_{11} = Var(t_3') = 1/36$ and $E(l_3) = E(t_2') + E(t_3') = 4/6$. All the other α_i's and β_{ij}'s can be computed similarly and we obtain

$$\boldsymbol{\alpha} = \frac{1}{6}\begin{pmatrix} 1 \\ 1 \\ 4 \\ 3 \end{pmatrix}, \quad \boldsymbol{\beta} = \frac{1}{36}\begin{pmatrix} 1 & 1 & 1 & 0 \\ 1 & 1 & 1 & 0 \\ 1 & 1 & 10 & 9 \\ 0 & 0 & 9 & 9 \end{pmatrix}$$

We now consider how a minimum variance estimator of θ can be derived from **m**. Notice that **m** can be expressed as a general linear model

$$\mathbf{m} = \theta\boldsymbol{\alpha} + \boldsymbol{\epsilon} \tag{30}$$

where the error term is $\boldsymbol{\epsilon} = \mathbf{m} - \theta\boldsymbol{\alpha}$ whose variance matrix is given by

$$Var(\boldsymbol{\epsilon}) = \theta\boldsymbol{\gamma} + \theta^2\boldsymbol{\beta}$$

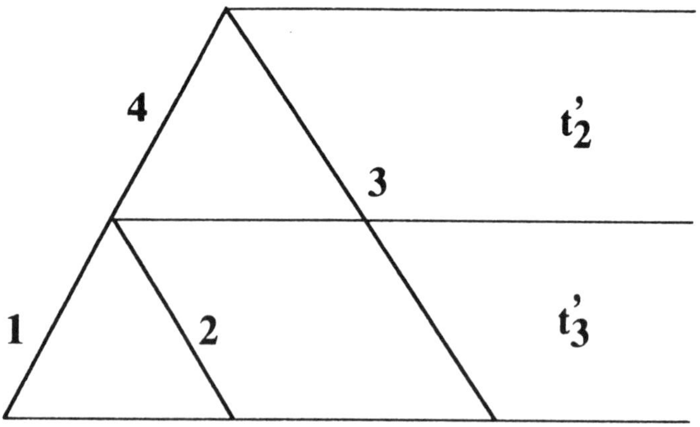

Figure 4. An example of genealogy of three genes.

From the theory of general linear model (for example, Searle[15]), we obtain the minimum variance linear unbiased estimator of θ as

$$\hat{\theta} = \left[\frac{\boldsymbol{\alpha}^T(\boldsymbol{\gamma} + \theta\boldsymbol{\beta})^{-1}}{\boldsymbol{\alpha}^T(\boldsymbol{\gamma} + \theta\,\boldsymbol{\beta})^{-1}\boldsymbol{\alpha}} \right] \mathbf{m} \tag{31}$$

However, this equation does not give us an estimate of θ directly because the computation of $\boldsymbol{\gamma} + \theta\boldsymbol{\beta}$ requires the value of θ, which is unknown. Therefore, an iteration procedure is necessary. Obviously, an estimate of θ can be obtained as the limit of the series:

$$\theta_{k+1} = \left[\frac{\boldsymbol{\alpha}^T(\boldsymbol{\gamma} + \theta_k\boldsymbol{\beta})^{-1}}{\boldsymbol{\alpha}^T(\boldsymbol{\gamma} + \theta_k\boldsymbol{\beta})^{-1}\boldsymbol{\alpha}} \right] \mathbf{m}$$

taking θ_0 to be an arbitary non-negative number. Denoting the limit of the series as $\hat{\theta}_m$, Fu[3] called $\hat{\theta}_{\mathrm{m}}$ the BLUE of θ (best linear unbiased estimator of θ) because (31) is a linear function of m $1,\ldots,m_{2(n-1)}$. Strictly speaking, $\hat{\theta}_m$ is not a linear function in $\mathbf{m}$ because of the iteration scheme.

To use $\hat{\theta}_m$ as an estimate of θ, the genealogy of a sample has to be inferred. Because an inferred genealogy often contains errors, the BLUE of θ based on an erroneous genealogy of a sample is likely to be biased. It is therefore important in practice to be able to correct the bias of estimation. Under the neutral Wright-Fisher model, the simplest treeing method, the unweighted pair-group method with arithmetic mean (UPGMA) seems to be an adequate method for reconstructing the genealogy of a sample (see e.g. Nei[10] for a detailed description of the method). To use UPGMA, one needs to calculate the number of mutations separating each pair of sequences (under infinite-site model, this number is the same as the number of nucleotide difference between the two sequences). These numbers form a distance matrix upon which the UPGMA is applied to obtain the genealogy of a sample. Fu[3] found that when the BLUE procedure is applied to the genealogy reconstructed by UPGMA, there is indeed, on average, a bias in the estimate of θ. The bias can be corrected using the following regression equation:

$$\hat{\theta} = \left(0.0335\sqrt{n-2} + 0.998\sqrt{\hat{\theta}_U} \right)^2 \cdot \tag{32}$$

where $\hat{\theta}_U$ is the BLUE of θ based on the genealogy constructed by UPGMA. The estimator $\hat{\theta}$ was found to be nearly unbiased and had a variance close to the minimum variance V_{min}. Therefore, the combination of BLUE of θ and UPGMA method results in a nearly optimal estimator of θ.

Detection of Natural Selection

Tajima[13] proposed to use the two different estimates $\hat{\theta}_T = \Pi_n$ and $\hat{\theta}_W = K/a_n$ of θ to detect selection. His test statistic is

$$T = \frac{\Pi_n - K/a_n}{\sqrt{Var(\Pi_n - K/a_n)}} \tag{33}$$

The rationale for this test is as follows. Since K ignores the frequency of mutants it is strongly affected by the existence of deleterious alleles, which are usually kept in low frequencies. In contrast, Π_n is not much affected by the existence of deleterious alleles because it considers the frequency of mutants. Thus if some of the sequences in the sample have selective effects, the estimate of θ based on K will be different from that based on Π_n. Therefore, the difference $\Pi_n - K/a_n$ can be used to detect the presence of selection. The denominator is intended to normalize the test. Tajima suggested the use of a beta distribution as an approximation of the distribution of the test so that critical values of the test can be obtained, but Monte-Carlo simulation would be a more accurate method for obtaining the critical values of the test.

One can construct a test of the same type as T for any pair of estimates of θ as long as the variance of their difference can be calculated. However, in order for such a test to be useful, the two estimates must be sufficiently different when selection is present. Because the expectations of η_e and $\eta_i/(a_n - 1)$ are both equal to θ under the neutral Wright-Fisher model and because they are likely to be different when selection is present, Fu and Li[4] proposed the following test:

$$D = \frac{\eta_e - \eta_i/(a_n - 1)}{\sqrt{Var[\eta_e - \eta_i/(a_n - 1)]}} \tag{34}$$

When deleterious mutations are frequent and purifying selection is in action, most of the deleterious mutants will be eliminated from the population and those which are around at the time of sampling are most likely to have arisen recently so that there is not enough time for them to be eliminated by selection. Recent mutations are close to the tips (external nodes) in the genealogy and therefore are mostly included in the value of η_e. In contrast, mutations in the internal branches are most likely to be neutral and $\eta_i/(a_n - 1)$ is not strongly affected by the presence of selection. Therefore, these two estimates of θ should be useful for testing the presence of selection. An advantage of D over T is that η_e and η_i are weakly correlated whereas Π_n and $\hat{\theta}_W$ are strongly correlated (Figure 2).

Because Π_n is less affected than η_e by the presence of selection, an obvious variant of test D is

$$F = \frac{\Pi_n - \eta_e}{\sqrt{Var[\Pi_n - \eta_e]}} \tag{35}$$

The variances of $\eta_e - \eta_i/(a_n - 1)$ and $\Pi_n - \eta_e$ are given by Fu and Li.[4] Note that since η_e is much more strongly affected than η by the presence of selection, we expect that F is at least as powerful as Tajima's test T for detecting the presence of selection.

The computation of the two tests, D and F, requires the number of external mutations for a given sample of size n. Under the infinite-site model, each segregating site is characterized by the frequencies of the two segregating nucleotides. A segregating site is said to have the pattern $i/(n - i)$ if one of the segregating nucleotides is found in i sequences and another nucleotide in n-i sequences. It is easy to see that each external mutation must correspond to a segregating site of pattern $1/(n - 1)$, which we call a singleton, but the reverse is not necessarily true. For example, every mutation on branch 4 of the tree in Figure 3 results in a segregating site of pattern $1/(n - 1)$ (where n=3) but it is by definition an internal mutation. Therefore, using the number of singletons in a sample as the value of η e can overestimate the true value. This difficulty can easily be overcome if an outgroup sequence is available, because an external mutation must be a singleton in the sample with and without the outgroup sequence. When an outgroup sequence is not available, a rough reconstruction of the genealogy of a sample is needed to infer the value of η_e. In the latter case it is more convenient to consider the number, η_s, of singleton segregating sites instead of the number of external mutations. The expectation of η_s can be shown[4] to be $[n/(n - 1)]\theta$. The analogous tests to D and F using η_s instead of η_e are

$$D^* = \frac{\frac{n}{n-1}\eta - a_n\eta_s}{\sqrt{Var\left[\frac{n}{n-1}\eta - a_n\eta_s\right]}} \tag{36}$$

$$F^* = \frac{\Pi_n - \frac{n-1}{n}\eta_s}{\sqrt{Var\left[\Pi_n - \frac{n-1}{n}\eta_s\right]}} \tag{37}$$

Critical values of tests D, F, D^* and F^* have been obtained by Monte-Carlo simulations and are given in Fu and Li.[4] At present, it is not known which one of these tests is the most powerful.

The Future

The development of statistical methods for studying DNA polymorphism is attracting more and more researchers because such data are becoming widely available. The cornerstone of such statistical methods is the coalescent theory of population genetics, which is being developed since early 1980s. Progress has been made in the last few years on the estimation of θ and the test of natural selection but much remains to be done.

An alternative to the mimimum variance estimator of θ described here is a Monte-Carlo maximum likelihood estimator by Felsenstein.[1] However, since his method requires a large amount of computation, its properties are unknown. Maximum likelihood based estimators appear to be more promising if it is constructed along the line of Strobeck[10] and Griffiths,[6] but estimators based on maximum likelihood cannot be significantly better than the minimum variance estimator of θ because the latter has been shown to be nearly optimal. However, significant progress in the estimation of θ can be expected under more sophisticated population models other than the neutral Wright-Fisher model. Developing efficient methods for estimating θ under alternative models is important because the neutral Wright-Fisher model is likely to be violated to some degree for many natural populations by the presence of recombination, selection, change in effective population size, population subdivision or a combination of them.

Developing more powerful statistical tests for testing the neutral Wright-Fisher model against unspecified alternatives is not easy because the distributions of quantities that may be useful for constructing tests are unknown and one often has to be content with first and second moments of these quantities, which are not easy to obtain either. However, greater progress can be expected in the near future on methods for testing the neutral Wright-Fisher model against a specific alternative hypothesis, such as certain types of selection, population growth and population subdivision. This is because such a test can be constructed as the likelihood ratio and methods to compute likelihood values under various population models are beginning to emerge.

References

1. J. Felsenstein. (1992b). Estimating effective population size from samples of sequences: a bootstrap monte carlo integration method. *Genetical Research* **60,** 209–220.

2. J. Felsenstein. (1992a). Estimating effective population size from samples of sequences: inefficiency of pairwise and segregation sites as compared to phylogenetic estimates. *Genetical Research* **56,** 139–147.

3. Y. X. Fu. (1994). A phylogenetic estimator of effective population size or mutation rate. *Genetics* **136,** 685–692.

4. Y. X. Fu, and W.-H. Li. (1993a). Maximum likelihood estimation of population parameters. *Genetics* **134,** 1261–1270.

5. Y. X. Fu, and W.-H. Li. (1993b). Statistical tests of neutrality of mutations. *Genetics* **133,** 693–709.

6. R. C. Griffiths. (1989). Genealogical tree probabilities in the infinitely-many-site model. *Journal of mathematical biology* **27,** 667–680.

7. R. R. Hudson. (1982). Testing the constant-rate neutral allele model with protein sequence data. *Evolution* **37,** 203–217.

8. R. R. Hudson. (1991). Gene genealogies and the coalescent process. In *Oxford Surveys in Evolutionary Biology,* Ed. by D. Futuyma and J. Antonovics, **7,** 1–44.

9. J. F. C. Kingman. (1982). On the genealogy of large populations. *J.Applied Probability* **19A,** 27–43.

10. M. Nei. (1987). *Molecular Evolutionary Genetics.* Columbia University Press.

11. C. Strobeck. (1983). Estimation of the neutral mutation rate in a finite population from DNA sequence data. *Theor. Pop. Biol.* **24,** 160–172.

12. F. Tajima. (1983). Evolutionary relationship of DNA sequences in finite populations. *Genetics* **105,** 437–460.

13. F. Tajima. (1989). Statistical method for testing the neutral mutation hypothesis by DNA popymorphism. *Genetics* **123,** 585–595.

14. G. A. Watterson. (1975). On the number of segregation sites. *Theor. Pop. Biol.* **7,** 256–276.

15. S. R. Searle. (1982). *Matrix Algebra Useful for Statistics.* John Wiley & Sons, New York.

11

Using Maximum Likelihood to Infer Selection from Phylogenies

Brian Golding

Abstract. *A maximum likelihood method to detect selection from phylogenies is reviewed. Models with k alleles and with unequal rates of mutation between alleles can be accommodated. The basis for the method is that unusually conserved sequences are an indication of functional importance. This feature can be used to infer the presence of purifying selection. The method is illustrated to search for evidence of selection acting on rRNA secondary structure. Sequence data from 51 species is used to examine conservation of rRNA secondary structures. These secondary structures are thought to be selectively important for the function of rRNA molecules. Strong selection for a specific secondary structure implies that nucleotide pairs that do not contribute to this structure will be rare. The frequency of different pairs of nucleotides at each site in the secondary structure is tabulated and the algorithm is used to obtain a maximum likelihood estimate of selection strength. A map of the potential strength of selection throughout the molecule is discussed.*

Introduction

A common and basic notion of most biologists is that conservation of sequence over many million years of evolution implies some level of purifying selection. It has been known since the early work of Zuckerkandl and Pauling[23] that some amino acids are more likely to be altered during evolution than others. They interpreted this as due to different selective constraints acting on the molecules. Similarly, it is well known that functionally more important proteins will evolve at a slower rate than less important proteins.[16] Indeed, this can even be observed within different regions of the same protein.

This concept of a connection between conservation and selection has been used to search DNA sequences for functional regions. For example, TATAA boxes can be identified by the consistent placement of this sequence upstream of some eucarya genes. It is also known that amino acid sequences change more slowly than DNA sequences, that the secondary structure of proteins changes more slowly than its primary sequence and that the three dimensional structure changes even more slowly than the secondary structure. Hence, ever more distant relationships between proteins can be detected through these different levels of conservation.

This notion is also used to aid the reconstruction of phylogenies. In order to study closely related species, characters or sequences that are not under strong influence of selection (e.g., the D-loop regions of mtDNA) should be chosen. These sites will not be conserved. Hence, they can change quickly and will provide more data that differentiates between closely related organisms. On the other hand, the phylogenetic relationships of distantly related species requires sequences that change slowly and will retain traces of their ancestry. These sequences are strongly influenced by selection and because of this sequence conservation, are more informative about distant relationships.[11]

A method has been developed[6] to translate the degree of conservation of alleles into a maximum likelihood measure of the strength of potential selection. I have investigated some properties of the method in Golding[7] and extended the model to consider an arbitrary k allelic states in Golding.[8] My purpose here is to further illustrate the method and to indicate how the algorithm might be applied. This method has been applied to intraspecific data in Golding[8] even though it is strictly applicable only to interspecific data. It's use will be shown here only to indicate another methodology to detect the effects of natural selection.

The example used here is the level of conservation in the secondary structures of rRNAs and this is translated into the corresponding strength of selection for a very simple model of secondary structure evolution. The data analyzed is from the structure of the 16/18S rRNAs of 51 species chosen from the three primary lines of descent of life; bacteria, eucarya, and archaea. These data do not provide new information about the structure of rRNA and it is assumed *a priori* that many paired positions are subject to some level of selection but the use of the algorithm provides some insight into the magnitude of selection that is implied by the degree of sequence conservation.

Theory

The concept underlying the general algorithm is that conservation of structure over long time periods implies some selective importance. When selection is very strong, a deleterious allele will rarely be observed. If selection is weak or absent, then nucleotide substitutions will be more common and the observation of deleterious alleles becomes more common. We wish to quantify the level of conservation and transform this into a maximum likelihood estimate of the selection coefficient. In addition to data on sequence conservation, this method requires some knowledge of mutation rates and of the phylogeny relating different species.

For example, if a character is not capable of mutating, then it will be conserved even in the absence of selection. It is also important to distinguish between mutation and substitution. Similarly, conservation of sequence may be due to close or recent ancestry rather than due to selection. Both of these possibilities must be eliminated.

An algorithm that does this was developed by Golding and Felsenstein.[6] This algorithm uses a model appropriate for sequences with two allelic states and with equal mutation rates between each allele. The likelihood of any state (or allele) must be determined for every site in the sequence and for every species and node within the phylogeny. Fortunately each site in the sequence can be considered independent of other sites. For one site, the likelihoods can be found recursively starting at the tips of the phylogeny. Following the notation of Golding and Felsenstein,[6] we can define the likelihood of part of an evolutionary tree subtended by the m-th node for a site with state i as $L^{(m)}_i$. Define P^t_{ij} as the probability that a site would change from state (or allele) i to state j within t generations.

As an example, assume that node m = 3 and that this node gave rise to two descendant nodes or taxa #1 and #2. Let the time separating node/species #3 and node/species #1 be t and between nodes/species #3 and #2 be t'. With these definitions the likelihood of an evolutionary tree can be found from

$$L_i^{(3)} = \left(\sum_{j=1}^{2} P_{ij}^{t} L_j^{(1)}\right)\left(\sum_{k=1}^{2} P_{ik}^{t'} L_k^{(2)}\right)$$

(Felsenstein[4]). The terms $L^{(1)}_j$, $L^{(2)}_k$ designate the likelihoods of that part of the evolutionary tree subtended by nodes #1 and #2, for sites with states j and k. If nodes #1 and #2 designate extant species (tips of the phylogeny) then these likelihoods are either 1 or 0 depending on whether the species does or does not have that state (allele) at that particular sequence site. The likelihoods for all interior nodes of a phylogeny can be calculated by recursively using this formula. The likelihood for the complete phylogeny is found stepwise beginning at the tips of the phylogeny and moving down the tree one node at a time. Each successive step uses the likelihoods just calculated such that the value determined for $L^{(3)}_i$ is used to find the likelihood of the next node. The likelihood of every subtree of every state is calculated for every node using the likelihoods calculated for the previous nodes. This continues until the root of the tree is reached and then the overall likelihood of the phylogeny is found by summing the products of the root likelihoods with the prior probabilities of each state. Without any further information, the prior probabilities of each state are usually taken to be their equilibrium frequencies.

It remains to determine P^t_{ij} for a model that incorporates selection. Assume that the mutation rate from state A to a is μ_1 and that the mutation rate from state a to A is μ_2. Selection on these states is genic selection with fitnesses of

1 for state A and 1-s for state a. The selection is the same within each species and operates at a constant strength over time. Each species consists of an effective population size of N individuals and the strength of selection is measured as the product Ns.

To calculate P^t_{ij} it is necessary to know the expected allele frequencies in a finite population with both selection and mutation acting. Unfortunately, the values of P^t_{ij} cannot be calculated exactly and hence the model is solved using two approximations. One approximation is appropriate when selection is weak and the other is appropriate when mutation is weak relative to selection.

The solution to P^t_{ij} when selection is weak follows the approach taken by Kimura[14] and Avery.[1] Let x designate the frequency of allele A, and E(x) its expectation. Consider a population that has only small changes in its allele frequency in any one generation. Then the expected frequency of A can be described by a series expansion in (Ns), where N is the effective size of the population, as

$$\begin{aligned} E(x)' &= \mu_2 + (1-\mu_1-\mu_2)E(x) + sE(x)|_{s=0} - sE(x^2)|_{s=0} \\ E(x)|_{s=0}' &= \mu_2 + (1-\mu_1-\mu_2)E(x)|_{s=0} \\ E(x^2)|_{s=0}' &= (2\mu_2+1/2N)E(x)|_{s=0} + (1-2\mu_1-2\mu_2-1/2N)E(x^2)|_{s=0} \end{aligned}$$

These equations provide an approximation to E(x) when Ns is small. They are a linear approximation in (Ns) although higher order approximations could also be determined if desired.

This system can be solved to give

$$\begin{aligned} E^t(x) = {} & \lambda_1{}^t E^0(x) + 4Ns(2+2\theta_2)(\lambda_1{}^t-\lambda_2{}^t)/(2+\theta_1+\theta_2)^2 \; E^0(x)|_{s=0} \\ & + st(\theta_1-\theta_2)\lambda_1{}^{t-1}/(2+\theta_1+\theta_2) \; E^0(x)|_{s=0} \\ & - 4Ns(\lambda_1{}^t-\lambda_2{}^t)/(2+\theta_1+\theta_2) \; E^0(x^2)|_{s=0} \\ & + \theta_2(1-\lambda_1{}^t)/(\theta_1+\theta_2) \\ & + 4Ns\theta_2(\theta_1-\theta_2)\,(1+(t-1)\lambda_1{}^t-t\lambda_1{}^{t-1})/(\theta_1+\theta_2)^2(2+\theta_1+\theta_2) \\ & + 8Ns\theta_2(1+\theta_2)\,(1-\lambda_1{}^t)/(\theta_1+\theta_2)\,(1+\theta_1+\theta_2)\,(2+\theta_1+\theta_2)^2 \\ & + 4Ns\theta_2(1+\theta_2)\,(1+\lambda_2{}^t)/(1+\theta_1+\theta_2)\,(2+\theta_1+\theta_2)^2 \\ & - 8Ns\theta_2(1+\theta_2)\lambda_1{}^t/(1+\theta_1+\theta_2)\,(2+\theta_1+\theta_2)^2 \end{aligned}$$

where $\theta_1=4N\mu_1$, $\theta_2=4N\mu_2$, $\lambda_1=(1-\mu_1-\mu_2)$, and $\lambda_2=(1-2\mu_1-2\mu_2-1/2N)$. This formula can be used to determine the probabilities of transition between allelic state as $P^t_{AA}=E^t(x)$ and $P^t_{Aa}=1-E^t(x)$ when $E^0(x)=1$ and $P^t_{aA}=E^t(x)$ and $P^t_{aa}=1-E^t(x)$ when $E^0(x)=0$.

The above approximation is only valid when Ns is small but another approximation can be used when Ns is large. This approximation assumes that mutation is weak relative to selection and hence selective fixations occur rapidly relative to the length of time between fixations. Great simplifications result if the fixation time is assumed to approach zero. In particular the probability of transitions can be simply determined from the probability of fixation of any particular allele and from the probability that alleles mutate. The probability of eventual fixation for an advantageous allele is

$$U(s,Ns) = (1-e^{-2s}) / (1-e^{-4Ns})$$

(Kimura[15]). Similarly the probability of eventual fixation for a deleterious allele is $U(-s,-Ns)$. There are $2N\mu_1$ new mutations to allele a and $2N\mu_2$ new mutations to allele A each generation. Using the above assumption, the probability of a change in the allelic state per unit time from A to a is $a = 2N\mu_1\, U(-s,-Ns)$ and the probability of a change from a to A is $\beta = 2N\mu_2\, U(s,Ns)$. Therefore the probabilities of change within t generations are

$$\begin{aligned} P^t_{Aa} &= \alpha/(\alpha+\beta)\,(1 - e^{-(\alpha+\beta)t}) \\ P^t_{aA} &= \beta/(\alpha+\beta)\,(1 - e^{-(\alpha+\beta)t}) \\ P^t_{aa} &= 1 - P^t_{aA} \\ P^t_{AA} &= 1 - P^t_{Aa}. \end{aligned}$$

The two approximations to the transition probabilities are quite different. The series approximation is only valid with weak selection and includes the chance of polymorphism. This approximation is used when $4Ns < 0.1$. The weak mutation approximation considers the relative proportion of species fixed for alternate alleles and since mutation is weak, does not permit extended polymorphism. This approximation is more appropriate when selection is strong and is used when $4Ns > 0.1$. Both approximations require that $N << t$.

Example

Ribosomal RNAs have been used to reconstruct the deepest branches of the phylogeny of life. These molecules are ubiquitous and, because of strong selection, their primary sequence changes slowly. This has permitted the most ancient splits within life to be quantified[2] and has demonstrated new, previously unrecognized branches of life.[22]

Often when the primary sequence of rRNA does change it is compensated by a nucleotide change elsewhere in the molecule. This phenomena of compensatory mutations was used to help determine the secondary structure of the rRNA molecule since these ancillary mutations were ones that retained a constant secondary structure.[9] It was suggested by Jukes[13] and Holmquist *et al.*[12] that the secondary structure itself may be under selection in addition to the primary sequence. This was empirically confirmed by Ozeki *et al.*[17] who showed that tRNAs with altered primary sequence but constant secondary structure were still viable. An excessive number of compensatory mutations has also been used to suggest the presence of some level of selection by Wheeler and Honeycutt.[20]

However, the presence of compensatory changes should not affect phylogenies.[18, 10]

The model of evolution described above is a simplication of the real world in that it assumes that parameters are constant over time (but the sequences need not be in equilibrium with respect to these parameters). The model assumes that mutation occurs between states at random without respect to the current state. For the purposes of this illustration, each state is assumed to be a pair of nucleotides and only two states are considered; a pair that cannot form hydrogen bonds and a pair that can. Hence the two states can be identified with a nucleotide pair in the rRNA molecule that either do or do not contribute to the secondary structure. We will consider not only classical Watson and Crick base pairs but also "G-U" and "U-G" pairs.

Only 6 of the 16 possible pairs of nucleotides contribute to the secondary structure (AU, UA, GC, CG, GU, UG). Assuming that mutations are random, the relative rates of mutation between states should be such that μ_1 is 62.5% and μ_2 is 37.5% of the total mutations. No consideration is made of the strength of the hydrogen bonding between nucleotides. The stability of bonding can change significantly depending on the actual nucleotide pair and on their neighbours.[19]

A site with nucleotide pairs that contribute to the secondary structure is assumed to be potentially advantageous to the species. A site with pairs that do not contribute to the secondary structure will be considered potentially deleterious to the species. Hence these two states correspond to states "A" and "a" in the previous section. The algorithm then asks a very simple question. Does an evolutionary model that incorporates selection for hydrogen bonded base pairs "*fit*" the data better than the same model without selection.

Data

The data consists of the primary sequence and secondary structures for the 16/18S rRNAs from 51 species. The species names are listed in Appendix 1 and include representatives from all major branches of life. The secondary structures for these species were determined by Gutell *et al.*[9] (see also Woese, Winker and Gutell[22]). Modelling of the 16S rRNA higher-order structure has utilized the comparative sequence approach. This method is based on the assessment that certain RNA sequence types will form a common higher-order structure. The first successful application of this method was performed on tRNA. It was realized in the early 1970s that different tRNA primary sequences were constrained to form a common higher-order structure. This assumption was borne out by various experimental approaches. Given this method's success with tRNA, it has more recently been applied to the growing number of RNA molecules, including but not limited to 5S, 16S and 23S RNA.

All of the species listed in the appendix are distantly related. The maximum sequence similarity is 93.2% between *Xenopus laevis* and man. But the divergence of mammals and amphibians is very ancient and this suggests that the actual phylogeny may not be a critical factor in this case. To check this, a phylogeny was reconstructed for these sequences using a consensus tree from 100 bootstrapped neighbor-joining trees. The branch lengths were chosen to minimize the squared deviation of observed and expected pairwise differences given a molecular clock.[5] The phylogeny used for the above algorithm is shown in Figure 1. Bootstrapped trees clearly separated the three major groups; archaea, eucarya and bacteria (each 100 times out of 100), but there are some other questionable placements of species. For example, *Blastocladiella* (12) is placed outside of the other fungi. The bar at the top of Figure 1 indicates the shortest branch length—approximately 340 million years separating mammals and amphibians. In a more general situation the phylogeny is of critical importance since different emphasis must be placed on shared characters depending on the phylogenetic relationships between the sequences. But when many different trees, in addition to that shown in Figure 1, were tested, all gave similar answers.

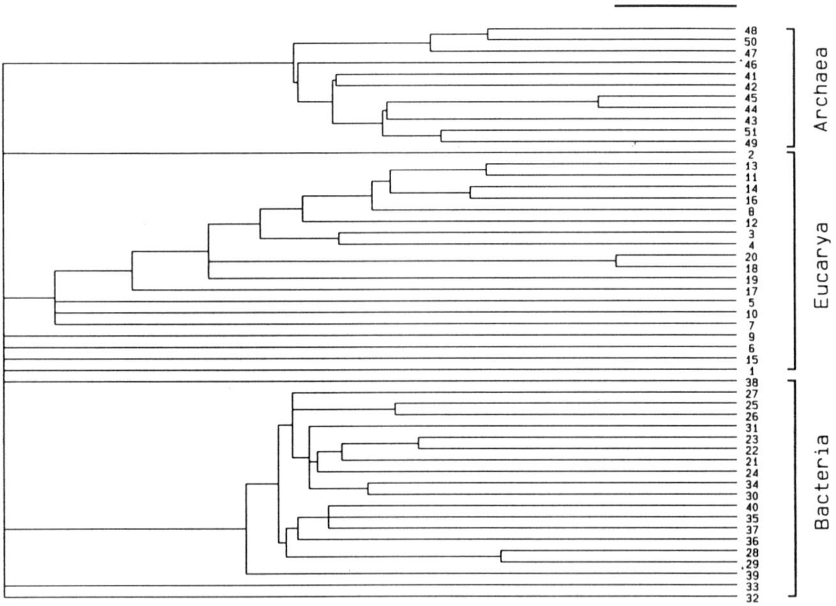

Figure 1. Phylogeny relating the 51 species listed in the appendix. The topology was determined from a consensus tree using one hundred bootstrapped neighbor-joining trees. Branch lengths minimize the squared deviations from observed and expected pairwise distances given a molecular clock (Felsenstein[5]). Numbers correspond to the species listed in the appendix.

Results & Discussion

In the total absence of selection one would expect to see any nucleotide pair at any one location in the secondary structure among the 51 species. Hence approximately 62.5% of species would be expected to have a location with nucleotide pairs that do not contribute to secondary structure and 37.5% would have pairs that do contribute. This is not what is observed; most species are restricted to pairs that can form strong hydrogen bonds. Table 1 shows the observed nucleotide pairs within the first two stems of the secondary structure. It shows that the expectation of only 37.5% is not met. While different species may have different nucleotide pairs, each pair is still usually able to form strong hydrogen bonds. Only one species has a single location with a pair that does not contribute to the secondary structure.

This conservation of paired nucleotides even when the individual nucleotides change suggests that selection may be acting on the secondary structure. Stronger selection would be implied by greater conservation. While we would like to estimate this selection, it is necessary to demonstrate that selection acting on individual nucleotides does not exist. (Individual nucleotides can be tested for the influence of selection ignoring all secondary structure using the method presented in Golding.[8]) The model being tested here examines only whether or not a particular nucleotide pair has one of two states; paired/unpaired. It then asks if selection improves the likelihood of the observation.

The likelihood of observing between 0 to 5 species without nucleotide pairs that contribute to the secondary structure from a total of 51 species is shown in Figure 2. If all 51 species have "correct" pairs, then the likelihood increases as the strength of selection increases. This continues indefinitely with the likelihood approaching one as 4Ns increases. This is because, if the alternate state is completely lethal, it is ensured that every species will have a "correct" pair. Hence, the likelihood of observing all species in this state increases up to one.

Table 1. Nucleotide pairs for the first two stems in the 16S RNA secondary structure.

	Helical Pairs						
Sites[a]	AU	UA	GU	GC	UG	CG	others
9/25	—	—	—	20	—	31	
10/24	20	20	1	—	—	10	
11/23	1	—	—	50	—	—	
12/22	1	—	—	28	22	—	
13/21	—	—	—	—	51	—	
17/918	—	50	—	—	—	—	1-UC
18/917	—	4	—	—	—	47	
19/916	1	2	—	—	—	48	

[*] Numbering scheme is that appropriate for *E. coli*.

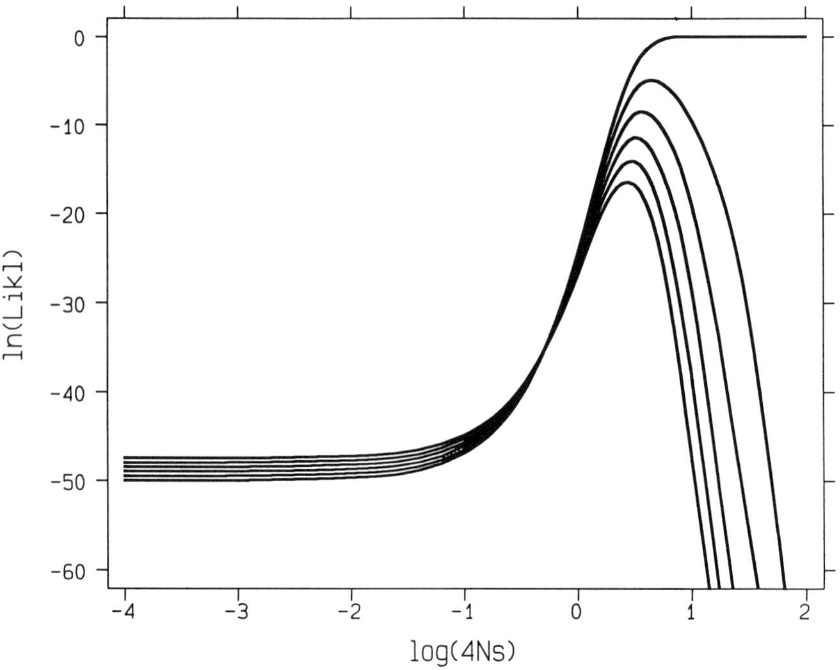

Figure 2. The likelihood plotted against selection (4Ns) with 0 to 5 species out of 51 species having nucleotide pairs that do not contribute to a secondary structure. The curves range 0 to 5 with 0 uppermost on the right and 5 lowermost on the right.

The likelihood when one of 51 species has nucleotide pairs that do not contribute to the secondary structure gives a different response to increasing selection. Initially, the likelihood of this observation again increases as selection increases. However, if selection becomes too strong then it becomes unlikely to observe even a single species with an "incorrect" pair. Therefore the likelihood first increases and then decreases as selection increases. The likelihood is maximized when 4Ns = 4.48. The maximum likelihood of 4Ns slowly decreases in size when more than one species has pairs that do not contribute to the secondary structure. In addition, the difference between the maximum likelihood and the likelihood at 4Ns = 0 also decreases.

Table 2 gives the maximum likelihood estimate of 4Ns with different degrees of conservation. It can be seen that the maximum likelihood value of 4Ns falls off quickly as the number of species with variant sequences initially increases, but then falls more slowly with larger numbers of species. In addition, a likelihood ratio test for the significance of selection can be done by comparing the relative size of the likelihoods in the presence or absence of selection. The probability that the increase in the likelihood is due to chance alone is shown in the last

Table 2. The likelihood of observing species with unpaired nucleotides without selection and the maximum likelihood estimate of 4Ns when selection occurs.

Number of species lacking H-bond	Ln Likelihood with no selection[a]	Maximum Ln Likelihood	MLE of 4Ns	Probability
0	−50.0191	0.0	<6.31>[b]	$< 10^{-10}$
1	−49.5084	−4.9062	4.48	$< 10^{-10}$
2	−48.9977	−8.4595	3.72	$< 10^{-10}$
3	−48.4870	−11.4128	3.28	$< 10^{-10}$
4	−47.9762	−14.0111	2.96	$< 10^{-10}$
5	−47.4655	−16.4093	2.72	$< 10^{-10}$
10	−44.9119	−25.2868	1.88	$< 10^{-8}$
15	−42.3582	−30.9103	1.34	$< 10^{-5}$
20	−39.8046	−34.1254	0.92	$< 10^{-3}$
25	−37.2510	−35.3340	0.53	0.05
30	−34.6973	−34.5533	0.15	0.41
35	−32.1437	−32.1437	0.0	1.00

[a] These were calculated assuming that $4Ns = 10^{-4}$.

[b] The MLE at #0 is arbitrarily set at log(4Ns)=0.8 to indicate a value after the curve in Figure 2 has leveled off—the true maximum is infinite.

column of Table 2. Only when the number of species with variant pairs becomes more than 25 does the evidence for selection become equivocal. By then, the value of 4Ns has decreased to 0.53. There can be strong selection (4Ns > 1) even with many species that do not have well-conserved locations. Thus studies of secondary structure that discount a potential pair due to a large number of non-compensatory mutations may be ignoring some real structure.

Collecting all the data yields a map of selection strength as shown in Figure 3. Here a vertical line has been added between each nucleotide pair that contributes to the secondary structure. The length of the vertical line corresponds to the maximum likelihood estimate of 4Ns for that particular nucleotide pair. The longer the line, the larger the value of 4Ns. A value of 4Ns=6.31 was chosen when the nucleotide pair is completely conserved among all 51 species. This value of 4Ns corresponds to the plateau of the likelihood surface in Figure 2. The dotted areas of the secondary structure are regions where the alignments/structures is deemed questionable due to sequence variation among the species. In these regions, alignment between non-homologous pairs would tend to inflate the estimates of 4Ns and to give overestimates in these regions.

The map in Figure 3 shows that many secondary structure regions of the rRNA molecule have a strong functional aspect in all species examined. Virtually all of the lines that are visible as lines rather than dots indicate significant evidence for selection. Interestingly, the strength appears to fall off in the more peripheral

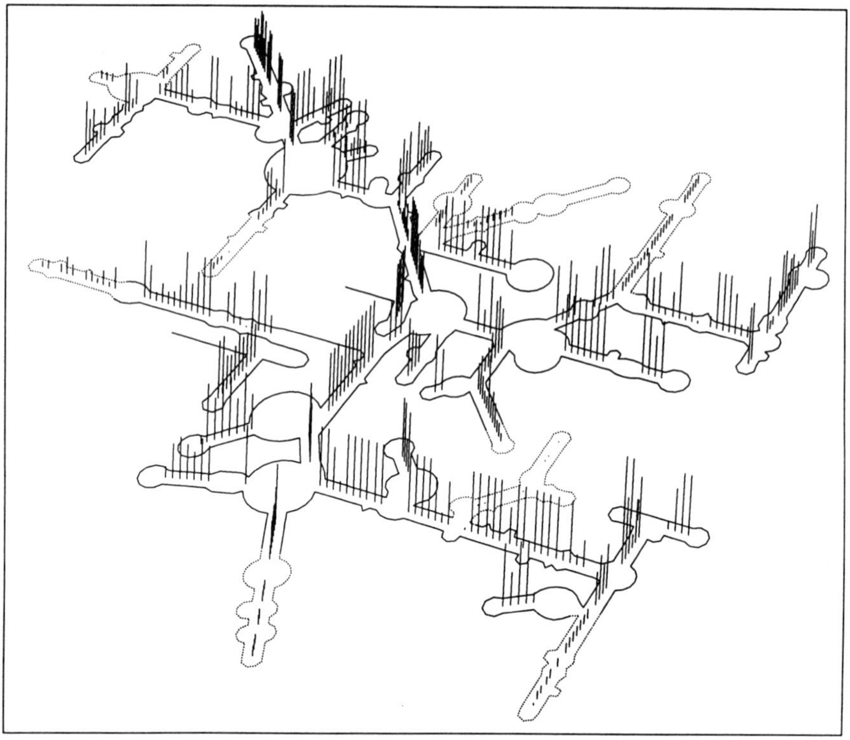

Figure 3. A map of the strength of selection on the rRNA molecule from 51 species. Stronger selection is indicated by longer vertical lines. The dotted lines are regions where the structure is variable or questionable—selection will be overestimated in these areas.

stems throughout the molecule. Central regions appear to be more highly conserved and hence under stronger selection. In addition, whole stem/loop structures appear to have similar levels of selection.

Some of the stem/loops that are implicated in binding such as in the upper left corner of Figure 3 are never-the-less not strongly selected for their secondary structure (relative to other regions of the molecule). Similarly, the two stem/loops at the 3' end of the molecule (middle left in Figure 3) are involved in subunit association[3] but are not strongly conserved. Other aspects of the molecular structure may be of greater importance here.

It should be stressed that this is an application of a simple model to a complicated situation. Selection may actually be stronger or weaker in some species but the use of the average degree of conservation to estimate selection will obscure this. Selection may be acting strictly upon individual nucleotide sites but one would then have to explain why so many paired sites covary. The algorithm provides a transformation of the level of sequence conservation to the more interesting and biological quantity of selection. It shows that the biological phenomena of strong natural selection may be occurring without the strict sequence conservation normally used to infer it. From Table 2 the transformation between these facets is not of a simple linear form. The breadth of the curves in Figure 2 gives an indication of the variance possible for the estimates of 4Ns and the likelihood approach permits the application of proven statistical tests such as the likelihood ratio test.

Acknowledgments

I would like to thank Robin Gutell for providing me access to this interesting data set and for helping me through it.

References

1. P. J. Avery. (1978). Selection effects in a model of two inter-migrating colonies of finite size. *Theor. Pop. Biol.* 13, 24–39.
2. R. Cedergren, M. W. Gray, Y. Abel and D. Sankoff. (1988). The evolutionary relationships among known life forms. *J. Mol. Evol.* 28, 98–112.
3. A. E. Dahlberg. (1989). The functional role of ribosomal RNA in protein synthesis. *Cell* 57, 525–529.
4. J. Felsenstein. (1981). Evolutionary trees from DNA sequences: A maximum likelihood approach. *J. Mol. Evol.* 17, 368–376.
5. J. Felsenstein. (1992). PHYLIP ver 3.4. Herbarium, University of California, Berkeley.
6. G. B. Golding and J. Felsenstein. (1990). A maximum likelihood approach to the detection of selection from a phylogeny. *J. Mol. Evol.* 31, 511–523.
7. G. B. Golding. (1993a). Estimating selection coefficients from the phylogenetic history. Pp 61–78 in *Mechanisms of Molecular Evolution*, ed by N. Takahata and A. G. Clark. Sinauer Associates, Inc., Sunderland Mass.
8. G. B. Golding. (1993b). Maximum likelihood estimates of selection coefficients from DNA sequence data. *Evolution*, in press.
9. R. R. Gutell, B. Weiser, C. R. Woese and H. F. Noller. (1985). Comparative anatomy of 16S-like ribosomal RNA. *Progress in Nuc. Acid Res. and Mol. Biol.* 32, 155–216.
10. S. B. Hedges, K. D. Moberg and L. R. Maxson. (1990). Tetrapod Phylogeny Inferred from 18S and 28S Ribosomal RNA Sequences and a Review of the Evidence for Amniote Relationships. *Mol. Biol. Evol.* 7, 607–633.
11. D. M. Hillis and C. Moritz. (1990). *Molecular Systematics*. Sinauer Associates Inc., Sunderland, Mass.
12. R. Holmquist, T. H. Jukes and S. Pangburns. (1973). Evolution of transfer RNA. *J. Mol. Biol.* 78, 91–116.

13. T. H. Jukes. (1969). Recent advances in studies of evolutionary relationships between proteins and nucleic acids. *Space Life Sci.* 1, 469–490.
14. M. Kimura. (1955). Solution of a process of random genetic drift with a continuous model. *Proc. Natl. Acad. Sci. USA* 41, 144–150.
15. M. Kimura. (1962). On the probability of fixation of mutant genes in a population. *Genetics* 4, 713–719.
16. M. Kimura. (1983). *The Neutral Theory of Molecular Evolution*. Cambridge University Press, New York.
17. H. Ozeki, H. Inokuchi, F. Yamao, M. Kodaira, H. Sakano, T. Ikemura, and Y. Shimura. (1980). Genetics of nonsense suppressor tRNAs in *Escherichia coli*. In *Transfer RNA: Biological Aspects*, ed. D. Soll, J. Abelson and P. Schimmel, pp. 341–362. Cold Spring Harbor Laboratory.
18. A. B. Smith. (1989). RNA sequence data in phylogenetic reconstruction: testing the limits of resolution. *Cladistics* 5, 321–344.
19. D. H. Turner and N. Sugimoto. (1988). RNA structure prediction. *Ann. Rev. Biophys. Biophys. Chem.* 17, 167–192.
20. W. C. Wheeler and R.L. Honeycutt. (1988). Paired sequence difference in ribosomal RNAs: Evolutionary and phylogenetic implications. *Mol. Biol. Evol.* 5, 90–96.
21. C. R. Woese and G.E. Fox. (1977). Phylogenetic structure of the prokaryotic domain: the primary kingdoms. *Proc. Natl. Acad. Sci. USA* 74, 5088.
22. C. R. Woese, S.Winker and R.R. Gutell. (1990). Architecture of ribosomal RNA: Constraints on the sequence of "tetra-loops". *Proc. Natl. Acad. Sci. USA* 87, 8467–8471.
23. E. Zuckerkandl and L. Pauling. (1965). Evolutionary divergence and convergence in proteins. In *Evolving Genes and Proteins*, ed. V. Bryson and H.J. Vogel, pp. 97–166. Academic Press, New York.

Appendix

Eucarya

1. Giardia lamblia
2. Vairimorpha necatrix
3. Tetrahymena thermophila
4. Paramecium tetraurelia
5. Plasmodium berghei
6. Trypanosoma brucei
7. Dictyostelium discoideum
8. Prorocentrum micans
9. Naegleria gruberi
10. Acanthameoba castellanii
11. Saccharomyces cerevisiae
12. Blastocladiella emersonii
13. Neurospora crassa
14. Chlamydomanas reinhardtii
15. Euglena gracilis
16. Zea mays
17. Caenorhabditis elegans
18. Xenopus laevis
19. Drosophila melanogaster
20. Human

Bacteria

21. Agrobacterium tumefaciens
22. Neisseria gonorrhoeae
23. Escherichia coli
24. Myxococcus xanthus
25. Bacillus subtilis
26. Clostridium innocuum
27. Mycoplasma gallisepticum
28. Streptomyces ambofaciens
29. Mycobacterium bovis
30. Heliobacterium chlorum
31. Leptospira interrogans
32. Bacteroides fragilis
33. Flavobacteria heparinum
34. Anacystis nidulans
35. Thermomicrobium roseum
36. Deinococcus radiodurans
37. Thermus thermophilus
38. Planctomyces staleyii
39. Chlamydia psittaci
40. Thermotoga maritima

Archaea

41. Halobacterium volcanii
42. Methanothrix soehngenii
43. Methanococcus vanniellii
44. Methanobacterium formicicum
45. Methanobacterium thermoautotrophicum
46. Thermoplasma acidophilum
47. Sulfolobus solfataricus
48. Thermoproteus tenax
49. Thermococcus celer
50. Thermophilum pendens
51. Archaeoglobus fulgidus

12

Gene Trees with Background Selection

Richard R. Hudson and Norman L. Kaplan

Abstract. *Consideration of the coalescent process is shown to be useful for analyzing neutral variation linked to loci at which deleterious variation is maintained by mutation-selection balance. Formulas for expected nucleotide diversity and the expected number of polymorphic sites in a sample are obtained. Simulations based on the coalescent process demonstrate that such background selection will rarely result in rejection of the neutral model using Tajima's D statistic. Models with recombination are also considered.*

Introduction

Charlesworth, Morgan and Charlesworth have recently shown that the level of neutral variation at a locus can be significantly reduced by selection at linked loci acting against deleterious alleles maintained by mutation.[1] They derived approximate formulas for the reduction in nucleotide diversity that results from such background selection against deleterious mutations. They also obtained formulas for the expected number of polymorphic sites in the entire population under this model. They showed that the effects of background selection may be significant for some regions of the *Drosophila melanogaster* genome. The purpose of this paper is to demonstrate how approximate formulas for the expected nucleotide diversity, and the expected number of polymorphic sites in **samples**, can be derived by consideration of the gene genealogy of sampled alleles. A model with recombination will also be analyzed.

Charlesworth, Morgan and Charlesworth also suggested that background selection might be detectable, at least in smaller populations with Tajima's test of neutrality based on D (Ref 1 and 2). Tajima's test will detect an excess of rare variants, which under some circumstances is expected under background selection models.[1] Simulations based on the coalescent process will be brought to bear on this issue.

The Model

Consider a neutral locus within which no recombination occurs, and with a neutral mutation rate μ per site per generation. Under a standard Wright-Fisher model[3], with diploid population size of N, the nucleotide diversity, π, has expected value given by

$$E(\pi) = 4N\mu \tag{1}$$

and the expectation of S_n, the number of polymorphic sites per base pair in a sample of size n is

$$E(S_n) = 4N\mu \sum_{j=1}^{n-1} 1/j \tag{2}$$

(Ref 4). These are equilibrium results for a neutral locus which is isolated, i.e., unlinked to loci at which natural selection is operating. We wish to consider a neutral locus linked to a collection of loci at which deleterious mutations occur. We will assume that this collection of loci, at which deleterious mutations can occur, are completely linked to each other, and hence can be considered together as a single locus which we will refer to as the "deleterious region." Initially, the neutral locus and the deleterious region will be assumed to be completely linked. Later we will allow recombination between the neutral locus and the deleterious region with the relationship of the neutral locus and the deleterious region as shown in Figure 1. Note that this assumption about recombination is different and probably less realistic than the model considered by Charlesworth *et al.*[1] which allows recombination between all the loci involved, with the neutral locus centrally located. Our model simplifies the mathematics and permits us to gain some insight into the role of recombination. When recombination rates are zero the two models are identical.

The total rate of mutation to deleterious alleles for the deleterious region is

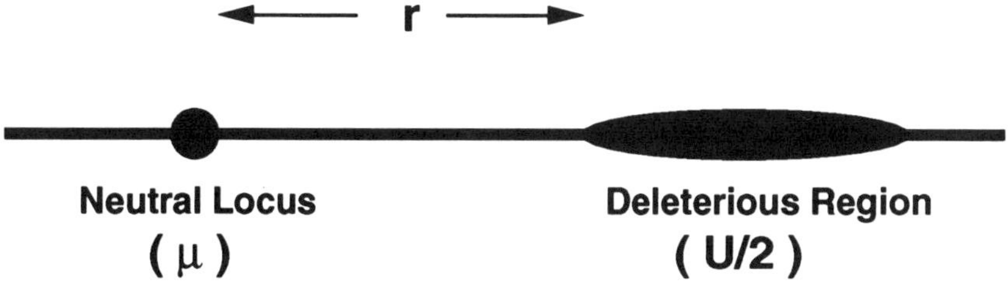

Figure 1. A simple recombination model, with a neutral locus linked to a deleterious region.

assumed to be U/2 per generation per gamete. (We follow the notation of Charlesworth *et al.*[1] as closely as possible.) The input of mutations will usually be assumed to be poisson distributed. Back mutation is assumed not to occur. Gametes that carry i deleterious mutations will be referred to as i-gametes. Let f_i denote the frequency in the population of i-gametes, and **f** denote the vector, ($f_0, f_1, f_2, f_3 \ldots$). In an infinite population, **f** will attain an equilibrium value referred to as mutation-selection balance. The values of **f** at equilibrium will depend on details of the selection model, including the strength of selection against the deleterious mutants, the dominance, the interaction between mutations, and the mutation rates. We will assume that such a mutation selection balance exists in our population and that **f** is absolutely constant. In fact, in any finite population **f** is expected to vary and for that reason our results must be considered approximate. For large populations the approximations will be quite accurate. Many of our results will be expressed as functions of **f**. These results are independent of the exact form of selection which resulted in a particular value of **f**. In some cases numerical results will be given for a particularly simple model (considered also by Charlesworth *et al.*[1]), in which all deleterious mutants have the same fitness affect and the interaction is multiplicative. Under this model, an individual heterozygous at i sites is assumed to have fitness $(1\text{-sh})^i$ relative to an individual with no deleterious mutations. This will be referred to as the multiplicative model. In this case, for autosomal loci the f_i are given by

$$f_i = \frac{\left(\frac{U}{2sh}\right)^i}{i!} e^{-(U/2sh)}, \tag{3}$$

(Ref 1).

We assume an equilibrium discrete generation model with mutation followed by selection as shown in Figure 2.

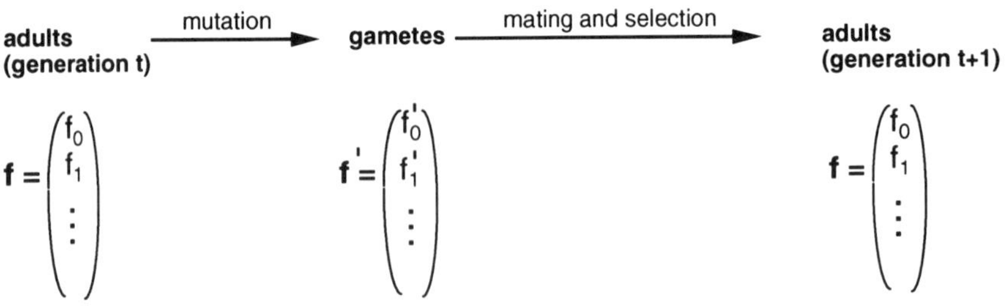

Figure 2. The life cycle assumed for the analysis of gene genealogies under background selection.

The Coalescent Process and Gene Genealogies

The expected nucleotide diversity due to neutral variation can be expressed as 2μ times the expected time (measured in generations) back to the most recent common ancestor of two randomly sampled copies of a gene[5]. Similarly, the expected number of polymorphic sites in a larger sample can be expressed as μ times the expected sum of the lengths of all the branches of the gene genealogy of the sample, again measured in generations[5]. In terms of equations, we have

$$E(\pi) = \mu T_2$$

and

$$E(S_n) = \mu T_n,$$

where T_2 is twice the expected time back to the common ancestor of two randomly sampled genes and T_n is the expected total length of the gene genealogy of a sample of size n. Under the Wright-Fisher neutral model T_2 is 4N generations, and hence the result (1). The expected total length of the gene genealogy of a sample of size n, is $4N\ \Sigma(1/j)$, from which the result (2) follows. To see how background selection affects nucleotide diversity and the number of polymorphic sites in samples we need only calculate T_2 and T_n, under the background selection model and compare them to the values under the neutral model. Sometimes the lengths will be measured in units of 2N generations, in which case under the neutral model, T_2 is 2, and T_n is $2\ \Sigma(1/j)$.

No Recombination

To calculate the mean size of a sample gene genealogy we procede in a fashion exactly analogous to that used to analyze gene genealogies under other models (Ref. 5 and references therein). That is, we follow the lineages of the sampled genes back in time, keeping track of changes in the ancestral genes and the occurrence of most recent common ancestors. The times between events are exponentially distributed with means which will be described below. An example genealogy is shown in Figure 3. As one traces back along a lineage of a gamete with one or more deleterious mutations, the lineage eventually jumps to less mutated classes, ultimately arriving at the 0-gamete class. (With recombination or back mutation, lineages can move to classes with larger numbers of mutations, too.) The two events which must be monitored in the history of a sample are coalescent events, in which two lineages have a common ancestor, and mutation events, in which a lineage moves from a more deleterious class to a less deleterious

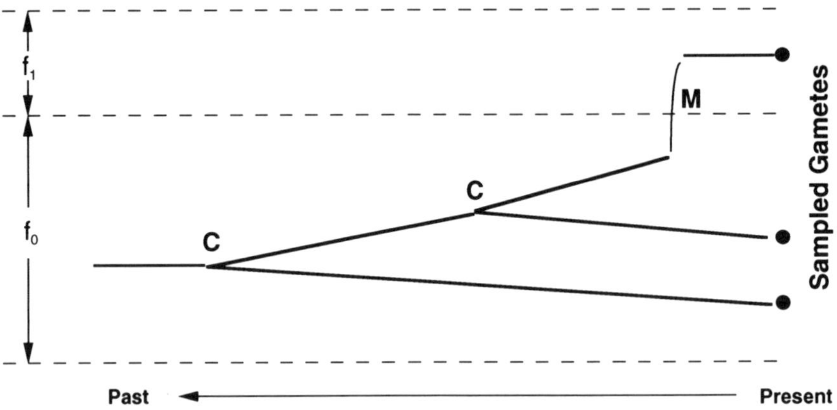

Figure 3. An example gene genealogy of a sample of three gametes. One of the sampled gametes is a 1-gamete and the other two are 0-gametes. A mutation event is marked "**M**" and two coalescent events are marked "**C**".

class (as we move back in time.) In many ways the coalescent process for this model is like the coalescent process under geographic structure models (or the balancing selection model), in which the different classes of gametes can be thought of as partially isolated subpopulations[5]. Coalescent events can only occur between members of the same type gamete (subpopulation) and mutation events are analogous to migration events.

We first consider a sample of one gamete, then a sample of two gametes and finally give results for an arbitrary size sample.

A 1-gamete, by definition carries one deleterious mutation. How long ago did the mutation in a randomly chosen 1-gamete occur? In other words, how far back in time must one go to find the first 0-gamete ancestor of a randomly sampled 1-gamete? The answer is easily expressed in terms of the probability P_{10}, that a randomly sampled 1-gamete in the current generation has a 0-gamete ancestor one generation back. It is easy to see that at equilibrium, the time back to first 0-gamete ancestor is approximately exponentially distributed with mean $1/P_{10}$. P_{10} is given by

$$P_{10} = \frac{f_0\left(\frac{U}{2}\right)e^{-U/2}}{f_0\left(\frac{U}{2}\right)e^{-U/2} + f_1 e^{-U/2}} = \frac{f_0\left(\frac{U}{2}\right)}{f_0\left(\frac{U}{2}\right) + f_1}. \tag{4}$$

This result is most easily seen by considering the population after mutation but before selection (see Figure 2). At this point of the life cycle, the frequency of 1-gametes, f_1', is given by the denominator of the middle part of (4). The frequency of 1-gametes that are newly mutated from 0-gametes is the numerator. The ratio is the frequency among 1-gametes of newly mutated 1-gametes. Selection does not distinguish between new and old 1-gametes, so the frequency of new 1-gametes among all 1-gametes is the same after selection as before, and therefore the ratio on the right hand side of (4) gives P_{10}.

More generally, the probability that an i-gamete is derived from a j-gamete one generation earlier is

$$P_{ij} = \begin{cases} \dfrac{f_j \dfrac{(U/2)^{i-j}}{(i-j)!}}{\displaystyle\sum_{k=0}^{i} f_k \dfrac{(U/2)^{i-k}}{(i-k)!}} & j<i \\ 0 & j\geq i \end{cases} \tag{5}$$

Unless U is quite large, P_{ij} is approximately zero, unless j is equal to i-1, in which case,

$$P_{i\,i-1} \approx \frac{f_{i-1}\,(U/2)}{f_{i-1}\,(U/2)+f_i}. \tag{6}$$

For the multiplicative model, using (3) and (6), one finds that $P_{i\,i-1} \approx ish$. So if sh equals 0.02, as suggested by Charlesworth *et al.*[1] for *Drosophila melanogaster*, a deleterious mutation on a 1-gamete is on average only 1/0.02 = 50 generations old.

Now we consider a sample of two gametes, and for simplicity suppose that the population has only 0-gametes and 1-gametes, so that $f_0 + f_1 = 1.0$. (This is only an approximation if the input of mutations is Poisson.) There are three possible sample configurations, namely, two 0-gametes, one 0-gamete and one 1-gamete, and two 1-gametes. Let T(2,0), T(1,1) and T(0,2) denote twice the mean time back to the common ancestor of two sampled gametes for these three sample configurations, respectively. Twice the overall mean time back to the common ancestor of two sampled gametes, T_2, is the average of these weighted by the frequency of the different sample configurations,

$$T_2 = f_0^2 T(2,0) + 2f_0 f_1 T(1,1) + f_1^2 T(0,2). \tag{7}$$

The probability that two 0-gametes have a common ancestor one generation back is $1/(2Nf_0)$, since there are $2Nf_0$ 0-gametes in the population one generation back, each of which is equally likely as an ancestor of each of the two sampled 0-gametes. This implies that

$$T(2,0) = 2(2Nf_0). \tag{8}$$

If the sample consists of one 0-gamete and one 1-gamete, then the time back to the common ancestor of the two gametes can be broken into two parts. The first part is the time back until the 1-gamete has a 0-gamete ancestor, at which point the ancestors of both sampled gametes are 0-gametes. The second part has mean T(2,0)/2, since it is just the additional time back until the two ancestral 0-gametes have a common ancestor. Thus,

$$T(1,1) = 2/P_{10} + T(2,0), \tag{9}$$

where P_{10} is given by (4).

If the sample consists of two 1-gametes, there are two distinct possibilities to consider. As one traces the two lineages back in time, the first event encountered might be a mutation, in which one of the 1-gametes has a 0-gamete ancestor, or the first event might be a common ancestor event in which the two 1-gametes have a common ancestor. (Here and in what follows, we ignore the possibility that more than one mutation, or a mutation and common ancestor event occur in the same generation.) The probability that one or the other of the two gametes has a 0-gamete ancestor each generation is approximately $2P_{10}$, and the probability of a common ancestor is approximately $1/2Nf_1$. The probability of one or the other event in any one generation is simply the sum, $2P_{10} + 1/2Nf_1$, so the mean time back until some event is $1/(2P_{10} + 1/2Nf_1)$. If the first event is a mutation, there is an additional time, with mean T(1,1)/2, back to the common ancestor of the sampled gametes. If the first event is a common ancestor event, there is no additional time in the gene genealogy. The probability that the first event is a mutation is $2P_{10}/(2P_{10} + 1/2Nf_1)$. Thus,

$$T(0,2) \approx 2/(2P_{10} + 1/2Nf_1) + \{2P_{10}/(2P_{10} + 1/2Nf_1)\}T(1,1). \tag{10}$$

Equations (7)–(10) together with (4) are sufficient to calculate T_2 as a function of f_0 and f_1, U, N and sh. If the multiplicative model applies, one can also use (3) (and assuming $f_1 \approx 1 - f_0$), T_2 can be calculated as a function of U, N and sh.

Note that for large population size, more precisely if $2Nshf_0 >> 1$ and $2Nshf_1 >> 1$, then (8)–(10) imply that $T(0,2) \approx T(1,1) \approx T(2,0)$ and hence that $T_2 \approx 4Nf_0$. That is, in a large population, all mutant gametes are very recent derivatives from 0-gametes. As one traces back the history of a pair of gametes, the ancestors of the sampled gametes very quickly become 0-gametes and then there is a relatively long time (with mean $2Nf_0$ generations) back to the common ancestor of the two gametes. The lengths of the lineages in which the ancestors are 1-gametes are negligible compared to the length of the lineages while the ancestors are 0-gametes. In this case, $E(\pi) = 4N\mu f_0$, which is just a factor of f_0 different from the neutral value, as was previously found[1].

Extension of these equations to a sample of n gametes with arbitrary vector **f** is straightforward. Consider a random sample of n gametes. The sample can be characterized by a vector $\mathbf{n} = (n_0, n_1, n_2, \ldots)$, where n_i is the number of i-gametes in the sample. (Note, that $\Sigma n_i = n$.) Under our model, **n** has a multinomial distribution with parameters n and **f**. We denote the mean genealogy size of a sample with configuration, **n**, by T(**n**). From now on time will be measured in units of 2N generations. Equations for T(**n**) will be given later. To obtain the overall mean genealogy size, T_n, one must average over these configurations using the multinomial probabilities as weights,

$$T_n = \sum_{n} M(\mathbf{n}|n,\mathbf{f})T(\mathbf{n}), \tag{11}$$

where the $M(\mathbf{n}|n, \mathbf{f})$ are multinomial probabilities referred to above and the summation is over all possible **n**, with $\Sigma n_i = n$. This is the generalization of (7).

Assuming that both mutation events and coalescent events are improbable in any single generation, so that the probability of more than one event occurring in a single generation is negligible, the mean time in the genealogy of a sample with configuration **n**, satisfies the following recursion approximately,

$$T(\mathbf{n}) \approx \frac{n}{X(\mathbf{n})} + \frac{1}{X(\mathbf{n})}\sum_{i=0} \frac{\binom{n_i}{2}}{f_i} T(\mathbf{n}-\mathbf{e}_i) + \frac{2N}{X(\mathbf{n})}\sum_{i=0}\sum_{j=0} n_i P_{ij} T(\mathbf{n}-\mathbf{e}_i + \mathbf{e}_j) \tag{12}$$

where

$$X(\mathbf{n}) = \sum_{i=0} \frac{\binom{n_i}{2}}{f_i} + 2N\sum_{i=0}\sum_{j=0} n_i P_{ij}$$

where the $\mathbf{e}_i$ are the unit vectors, (0, …,0,1,0,..,0), with a one in the ith position. If $n_i < 2$, $\binom{n_i}{2}$ is taken to be zero. The P_{ij} are given by (5). X(**n**) is the probability per 2N generations of some event occurring, either a common ancestor event or a mutation event. The first term on the right hand side of (12) is the mean time back to the first event times the number of lineages, n.

If $2NP_{10}f_i >> \binom{n}{2}$ then mutation events will be much more likely than coalescent events, until all ancestors are 0-gametes. (For the multiplicative model, this inequality is approximately equivalent to $2Nshf_i >> \binom{n}{2}$.) In other words, as one traces the genealogy back in time, relatively quickly all lineages will have become 0-gamete lineages, and the coalescent process from that point back in time, is just like the coalescent process for an isolated neutral locus, except the population size is Nf_0 instead of N. The genealogies in this case are distributed just like the isolated neutral case, except for a change in time scale by a factor of f_0. Thus all the T_i's will be reduced on average by a factor of f_0 from the standard isolated neutral case. However if the sample size is sufficiently large, then the time spent as deleterious gametes becomes significant relative to the times between coalescent events in the un-mutated class of gametes. In this case, the genealogies are not shortened as much as for smaller sample sizes. This is why there is a tendency for there to be an excess of polymorphic sites in large samples, relative to the nucleotide diversity observed. This effect will be more apparent in large samples from small populations, and when f_0 is small.

These equations can be used to examine the effect of background selection on the mean of π and S_n, however, to obtain higher moments or other statistical properties, simulations may be the most efficient method. Simulations based on the approximate coalescent process, which was described above, were carried out. These simulations were based on the same assumptions as (12), which permit one to quickly generate the genealogies of samples. Once genealogies are generated by Monte Carlo methods on the computer, one can address many questions about the neutral variation that these gene genealogies imply. It should be emphasized that these simulations cannot be used to check the accuracy of the above equations for π and S_n (or T_n), because the simulations are based on the same approximations as the equations. However, the simulations can be useful in cases where there is good reason to believe the approximations are excellent, but there are other statistical properties of the samples for which equations such as (12) are not available. We will use simulations to address the question: How often, under the background selection model, would one reject the neutral model because Tajima's D statistic is too small?

Results of the simulations are shown in Table 1, where one can examine the effects of sample size, population size and U, on the mean of D, and on the probability that D is less that D_{crit}, which is the approximate 0.025 critical value of Tajima's D statistic. The selection parameter, sh, is 0.02 in all cases. With the mutation rates and selection coefficients that we have examined, we find that when the population size is 10^5 or greater, there is very little effect of background selection on the mean of D, or on the probability of rejecting neutrality, with samples of up to 100 gametes. For much smaller population sizes there can be a substantial effect on the mean of D and the probability of rejection, especially

Table 1. Simulation results showing the affect of background selection on Tajima's D statistic. (4Nu = 15, sh = 0.02).

n	U	N	π^a	S_n^a	D^b	$P(D<D_{crit})^c$
10	0.0 (neutral)	—	1.00	1.00	−0.06	0.019
	0.01	100	0.87	0.91	−0.22	0.023
		3200	0.79	0.79	−0.08	0.022
		10^5	0.79	0.78	−0.06	0.022
		10^6	0.78	0.78	−0.08	0.023
	0.1	100	0.52	0.58	−0.54	0.019
		3200	0.10	0.11	−0.43	0.021
		10^5	0.08	0.08	−0.05	0.008
		10^6	0.08	0.08	−0.04	0.008
30	0.0 (neutral)	—	1.00	1.00	−0.10	0.021
	0.01	100	0.88	0.94	−0.33	0.032
		3200	0.78	0.79	−0.14	0.024
		10^5	0.78	0.78	−0.09	0.024
		10^6	0.78	0.78	−0.11	0.024
	0.1	100	0.51	0.67	−0.85	0.053
		3200	0.10	0.14	−0.84	0.133
		10^5	0.08	0.08	−0.09	0.026
		10^6	0.08	0.08	−0.05	0.023
100	0.0 (neutral)	—	1.00	1.00	−0.10	0.026
	0.01	100	0.87	0.97	−0.38	0.035
		3200	0.78	0.80	−0.18	0.034
		10^5	0.79	0.78	−0.09	0.025
		10^6	0.78	0.78	−0.11	0.028
	0.1	100	0.51	0.75	−1.01	0.104
		3200	0.10	0.19	−1.31	0.376
		10^5	0.08	0.09	−0.15	0.030
		10^6	0.08	0.08	−0.06	0.023

[a] These are the average values in 10,000 simulated samples divided by the expected value under the neutral model.

[b] This is the mean value of Tajima's *D* statistic in 10,000 simulated samples.

[c] $P(D<D_{crit})$ is the fraction of 10,000 simulated samples with $D<D_{crit}$. D_{crit} is −1.75, −1.72, and −1.61 for n=10, 30, and 100, respectively.

for larger sample sizes. For larger U, and consequently smaller f_0, we would expect the effects of background selection on Tajima's D to extend to larger population sizes than we have observed here.

Recombination

If recombination occurs at rate r per generation per gamete between the neutral locus and the deleterious region, as indicated in Figure 1, extensions to (12) can be obtained to calculate T_n. In fact, equation (12) can be used with the only modification being that the P_{ij} are given by the following instead of (5):

$$P_{ij} = \begin{cases} \dfrac{f_j \dfrac{(U/2)^{i-j}}{(i-j)!}}{\sum_{k=0}^{i} f_k \dfrac{(U/2)^{i-k}}{(i-k)!}} + rf_j & j<i \\ rf_j & j \geq i \end{cases} \tag{13}$$

Equation (12), without recombination, could be considered a recursion that can be solved sequentially for samples with fewer and fewer mutations. But with recombination, it appears necessary to solve the entire set of equations as a system. In figure 4 is shown a plot of T_2 as a function of r, obtained by solving (12) with (13), and assuming that N=3200, sh=0.02 and that f_i is given by (3) for i<7, that $f_7 = 1 - f_0 - f_1 - f_2 - f_3 - f_4 - f_5 - f_6$ and $f_i = 0$ for i>7.

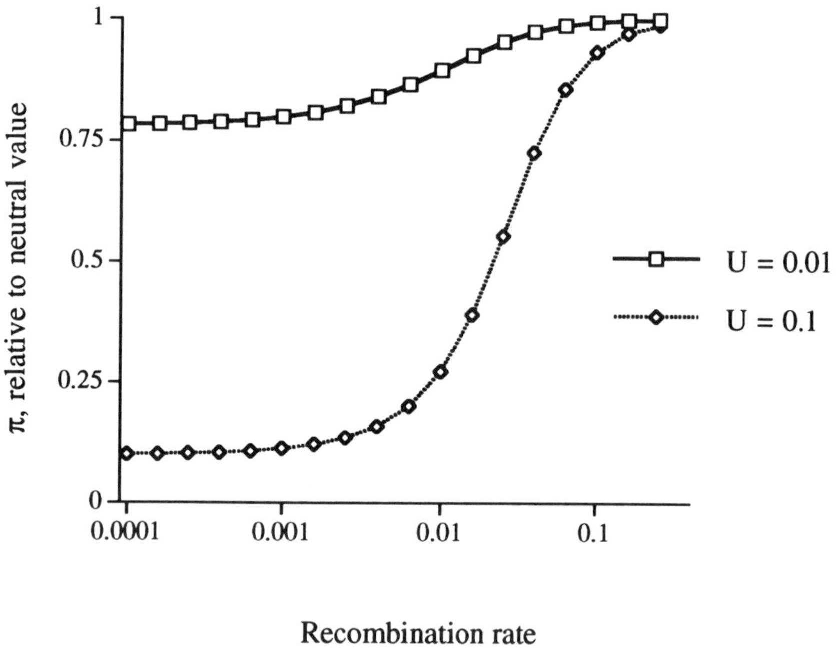

Figure 4. Expectation of π relative to the neutral expectation, plotted as a function of r, the recombination rate between the neutral locus and the deleterious region. These values were calculated with (12) and (13), assuming that N=3200, sh=0.02, and that f_i=0, for i>7.

We can get a better feeling for the dependence of T_2 on r by considering the multiplicative model, in which, as noted after equation (6), the relevant P_{ij} are simple functions of sh unless U is quite large. In this case, it is easy to see that the system (12) depends on the composite parameters 2Nsh, 2Nr and **f**, which is itself a function of U/(2sh). This has implications for simulations of this model aimed at the interpretation of variation in large natural populations. If the simulations are carried out with small populations, then it appears that the values of sh and r used should be scaled up, so that the values of Nsh and Nr are realistic. However, we have seen earlier, when considering the no recombination model, that as Nsh gets large, that properties of the gene genealogy become independent of Nsh, and become dependent on U/sh only. Thus, if the natural population of interest is large, so that Nsh is very large, then the simulation parameter, Nsh, needs to be large also, but not necessarily as large as the actual population being considered.

The apparent dependence of the system on 2Nr also needs to be considered. For the case, n=2 and when $f_0+f_1 \approx 1$, the system is small enough that an analytical solution can be obtained without great difficulty. However, the solution is cumbersome and will not be presented here. However, if 2Nsh is very large compared to one, the solution simplifies to :

$$T_2 = \frac{\left[1 + \left(\frac{r}{sh}\right)\right]^2 2f_0}{1+2\left(\frac{r}{sh}\right)f_0 + \left(\frac{r}{sh}\right)^2 f_0} \tag{14}$$

which shows that for large N, it is the ratio of r to sh that is important for determining the equilibrium level of variation. Again, it appears that simulations with realistic values of sh and r, but unrealistically small population size, can yield relevant results for interpreting variation in large natural populations. One does not need to use scaled up values of r, as long as the simulation value of Nr and Nsh are sufficiently large.

Using Equation (14), we see that if r<<sh, that T_2 is approximately $2f_0$, which is the no recombination result. Similarly, if r>>sh, T_2 is approximately equal to two, the result without any deleterious background mutation. Note that these are only approximate conditions since there is a dependence on f_0 as well. Figure 4 shows that these conditions are also appropriate at least in some cases when f_0+f_1 is not near one. This suggests that the more realistic recombination model of Charlesworth *et al.* might be approximated as follows. Perhaps, one can simply ignore the effects of deleterious loci that are too far away, say with r>sh, and treat all loci such that r<sh as if they were completely linked. That is, one uses the no-recombination model, but considers only loci such that r<sh. Recall in the Charlesworth *et al.* model, there are approximately 500 deleterious

loci on each side of the neutral loci. They asssume that sh=0.02 for their simulations. The loci most distant from the neutral loci recombine with the neutral loci approximately at rate 500c, where c is the rate of recombination between adjacent loci. Therefore if c is less than sh/500 = 0.02/500 = 4×10^{-5}, then the most distant loci still have $r < sh$, and we expect the no recombination model to apply. Indeed we see in Figure 1 of Charlesworth *et al.*[1] that for c less than 4×10^{-5}, the nucleotide diversity is roughly reduced by the factor $f_0 = \exp(-U/2sh)$, as with no recombination. For larger recombination rates, our approximation, suggests that the U in this expression should be replaced by a smaller number that corresponds to the fraction of loci within a recombination distance of sh of the neutral locus. The number of loci within that distance of the neutral loci is 2sh/c. Thus, U should be replaced (2sh/c)(U/M), where M is the total number of loci. Thus we expect the reduction in nucleotide diversity to be approximately

$$f_0 \approx \exp(-\{2shU/cM)\}/2sh\} = \exp(-U/cM) = \exp(-u), \qquad (15)$$

where u is U/cM, the deleterious mutation rate per diploid per recombination unit. The quantity f_0 is now to be interpreted as the frequency of gametes that are free of deleterious mutations within a recombinational distance of sh of the neutral locus of interest. Remarkably, this fraction, under the multiplicative model, appears to be approximately independent of sh, as long as the ends of the region are recombinationally sufficiently distant. Equation (15) should be regarded as tentative until simulations or analysis are carried out to verify its accuracy. We note only that (15) is roughly compatible with Figure 1 of Charlesworth *et al.*[1]

Conclusions

The coalescent approach appears promising for addressing some questions about the effects of background selection on linked neutral variation. Our analysis suggests that, under the multiplicative model, the effects of background selection may not depend on the selection coefficients, but instead on the total rate of production of deleterious mutants of strong effect per recombinational map distance. Tajima's D statistic appears to be relatively unaffected by background selection in large populations. These results are tentative, based on approximations with uncertain accuracy or on simulations with relatively few parameter combinations. Further work is required. In addition, it is important to examine non-multiplicative models and models with weaker selective effects.

Acknowledgments

This work was in part supported by U.S. Public Health Service grant GM42397.

References

1. C. Charlesworth, et al. (1993). *Genetics* 134, 1289–1303.
2. F. Tajima. (1989). *Genetics* 123, 585–595.
3. W. J. Ewens. (1979). *Mathematical Population Genetics*. Springer-Verlag, New York.
4. G. A. Watterson. (1975). *Theor. Pop. Biol.* 10, 256–276.
5. R. R. Hudson. (1990). in *Oxford Surveys in Evolutionary Biology*, edited by D. Futuyma and J. Antonovics, pp 1–41. Oxford University Press, Oxford.

13

Phylogenetic Analysis on the Edge: The Application of Cladistic Techniques at the Population Level

Robert DeSalle and Alfried P. Vogler

Abstract. *The potential of cladistic techniques in examining phylogeny at or below the population level is examined. Four methodologies for the reconstruction of relationships between populations (qualitative Hennigian analysis, gene frequencies in a maximum parsimony framework, population aggregation analysis and individuals as terminals) are summarized and their theoretical backgrounds explained. Each technique is placed into the context of Hennigian phylogenetic systematics and used to examine population level patterns in two groups of insects (Hawaiian* Drosophila *and* Cicindela dorsalis*). The "line of death" at the boundary of tokogeny and phylogeny is assessed in light of these techniques and the suggestion that phylogenetic analysis is not possible below this line is discussed.*

Anagenesis, Tokogeny and Phylogeny

The relationships of individuals within and among natural populations have been described largely using the concepts of classical population genetics. The Fisher-Wright model provided the predominant school of thought for this analysis with the emphasis on observations and theory concerning the change of allele frequencies with time. The major emphasis in this synthesis was on anagenetic change. Allele frequencies could be considered the currency of population genetics and the characterization of subdivision, gene flow and variability could be considered the goals. More recently, genealogical relationships within and between populations have become important subjects. Due to the fine detail that molecular genetic analysis can give to a study, more precise description of relationships within and between populations have become possible. Recent studies range from attempts to understand migration patterns and rates,[24] attempts to understand founder events,[15] and tests of "associations between phenotype and biochemical functions."[9] There is a need for the assessment of the theory and methodologies involved in analysis of population level data in a systematic framework.

Hennig[19] described phylogenetic analysis as a method primarily concerned with the retrieval of descent relationships among organisms. He also realized that there is a fundamental difference in the relationship of organisms in genetically separated lineages (phylogeny) and in interbreeding or recombining lineages (tokogeny). Figure 6 of his 1966 book, *Phylogenetic Systematics* has been reproduced in numerous publications and adequately demonstrates the "line of death"[41] between tokogeny and phylogeny. Complex reticulate patterns occur in populations of sexually reproducing organisms at the tokogenetic level, which is dominated by parent-offspring relationships. It is these reticulate tokogenetic relationships that are presumably not within the realm of phylogenetic inference.[19,23,31] These authors claim that if the tokogenetic "line of death" is crossed in phylogenetic analysis "lack of resolution is a likely outcome of a phylogenetic analysis of tokogenetic relationships among biparental organisms which are inherently non-hierarchical"[23] What we are left with in these kinds of studies are populations with gene frequencies on the one hand and individuals with attributes on the other. This paper attempts to describe the current methods and theory in systematics that attempt phylogenetic analysis at the boundary of tokogeny and phylogeny.

A Summary of Current Methods

Perhaps the easiest way to handle the treatment of data sets relevant to the tokogenetic-phylogenetic boundary is to take allele frequencies and convert them to some form of genetic distance.[27] The genetic distances are then used in a phenetic analysis to cluster the terminals on the basis of similarity. If on the other hand, the Hennigian definition of phylogeny is accepted, the notion of clustering on the basis of similarity is unacceptable.[4,18]

Several methods have been proposed to approach this dilemma in the framework of phylogenetic systematics. These methods range from not accepting a preconceived notion of species or population definition (treating individuals as terminals[41]) to strict population aggregation analysis.[11] In addition to these two extremes are several techniques such as the treatment of allele frequencies in a maximum parsimony framework.[35] Other methods exist such as transformation series analysis of polymorphic characters,[25] establishing consensus of polymorphic characters, qualitative Hennigian analysis[32] and assignment of plesiomorphic character states to polymorphic characters. Another approach is the coalescent theory.[9,17,20,21,22]

The Distinction Between Cladistic Analysis and Phylogenetic Analysis

Before describing the various approaches in detail it is essential to discuss the difference between phylogenetic analysis and cladistic analysis. A discussion of this distinction is necessary to establish what units are appropriate for phylogenetic analysis. The generation of hierarchic arrangements of relationships (trees) of terminal units (terminals) is the goal of cladistic analysis. Series of attributes of the terminals are scored and used under parsimony to search for such relationships. Cladistics is essentially a numerical procedure that makes no assumptions or invokes no models that relate the trees to biological processes.[11] Phylogenetic analysis on the other hand is "the reconstruction of descent relationships among terminals."[11] Although parsimony analysis can be used to infer these hierarchical relationships, this does not mean that cladistic analysis and phylogenetic analysis are equivalent. Certain assumptions are needed in the reconstruction of a descent history that are absolutely unnecessary for the implementation of cladistic analysis.

According to Davis and Nixon[11] two conditions of Hennig's model are essential for a cladistic analysis to be representative of phylogenetic relationships The first is that hierarchy must be present in the descent relationships among the terminals (or "taxa" being examined) for the model to work. The second condition is that all descendants of a common ancestor "retain all of the ancestor's characters, either in the original or transformed state." As an example of a system that fits these conditions, consider organellar DNA. Patterns of inheritance for this system fit both requirements and hence the hierarchy obtained using this attribute will be congruent with the descent hierarchy. Now consider sexually reproducing organisms. Such organisms usually do not fit the first condition, as any family genealogy most likely demonstrates reticulating patterns. However, even if the descent relationships of an attribute are hierarchical in sexually reproducing organisms, if the attribute is not fixed it is possible that it will not appear in any of the descendant populations (Figure 2 in Davis and Nixon[11]).

Inferring Trees from Frequency Data Using Maximum Parsimony

Qualitative Hennigian Method: The qualitative Hennigian method[26,32,33] relies on scoring alleles (or combinations of alleles) as present or absent regardless of frequency of their occurrence in a population. The initial step in this type of analysis consists of polarity assessment using an outgroup comparison with each allele being classified as ancestral, derived or unknown. The major problem with this method is the failure to detect the presence of rare alleles in small sample sizes as described by Patton and Avise[32] and Swofford and Berlocher.[35] Another problem is that optimal solutions using this method sometimes hypothesize ancestors that could not possibly have existed in the range of the original frequencies.[35]

Allele Frequency Maximum Parsimony[35]: This method uses direct allele frequencies and maximum parsimony to find the shortest tree. The algorithm they use divides the problem of tree construction into two sub problems. The first subproblem involves the assignment of allele frequencies to interior nodes in the tree so that tree length is minimal. The second subproblem involves finding the set of all possible trees that permit minimization of tree length. An exact solution for the first subproblem exists and a heuristic solution for the second is used in this approach.

This approach has been criticized by Crother[10] as the "some is better than none" approach. The basic criticism of Crother is that allele frequencies in a population can vary temporally. If this is the case then different phylogenetic scenarios for the same terminals are obtained for different time periods. Another criticism of this method applies in a similar way to continuous morphological characters. Operationally, continuous characters do not easily yield synapomorphic information, because of the difficulty of establishing character states along a continuum.[10] A final criticism against using allele frequencies as character states is that individuals of a species will rarely have the character state assigned to that species.[4] In fact, an individual will only have the same character state when the individual is a heterozygote for two alleles and the allele frequency of the taxon is 0.5.

Other Methods: Transformation Series Analysis (TSA) attempts to estimate the transformation series underlying a group of character states.[25,35] The ability to accomplish this obviates the major criticism of qualitative Hennigian analysis. A character state tree is constructed that represents the transformation series of the alleles present. The transformation series or character state tree represents the evolutionary relationships among the character states. In the case of allozymes it would "specify a hypothesis for the linear or branching pattern through which the various alleles at a locus have been derived from other alleles by mutation."[35] Such information is difficult to obtain using allozymes, but similar reasoning could be used with DNA based characters where determination of the transformation series or character state tree might be easier and more reliable due to the comparative precision of DNA sequencing technology. With this technique character states become much better defined and this definition eliminates the problem of meaningless synapomorphies that plagues qualitative Hennigian analysis.

Assignment of a character state on the basis of consensus (which allele is most frequent) has also been suggested as a method for dealing with polymorphism. This method is obviously unacceptable due to the arbitrary discarding of character state information. Assignment of the plesiomorphic character state or the designation of polymorphic attributes as missing have also been suggested as ways to deal with polymorphic attributes.[12] The practice of scoring polymorphic characters as missing data has been criticized[30] because cladogram length and consistency is altered by such scoring.

Population Aggregation Analysis

Characters, traits and phylogenetic species: The separation of phylogeny and tokogeny has played an important role in the development of the phylogenetic species concept. This species concept is based on the idea that organisms exhibit certain attributes that can be used to group or diagnose such organisms. Those groups of organisms (or populations) that can be distinguished from other such groups based on these attributes are phylogenetic species.[29] All individuals within these groups exhibit a transformed state for a given attribute which is not found in other populations. These attributes define the populations as diagnosable.[8] We follow the terminology of Nixon and Wheeler[31] and Davis and Nixon[11] who distinguish between "traits" and "characters." Traits are those attributes that distinguish individuals in a tokogenetic relationship while characters are attributes that distinguish or diagnose phylogenetically separate systems. These concepts are intricately tied to the phylogenetic species concept.[11,31]

It should be clear that reticulate relationships between or within populations effectively prevent the identification or discovery of characters that diagnose populations. A prerequisite for the origination or establishment of the phylogenetic species is therefore a non-reticulate pattern of relationships. Consequently, the "line of death" separating cladogenesis and tokogenesis is also the line that defines the level above which phylogenetic species can be recognized.

The population aggregation analysis method: Population aggregation analysis is a method by which "traits" can be discerned from "characters." The method uses variation patterns within populations to determine if an attribute is a trait or a character. The assignment of individuals to populations is also a critical aspect of this analysis and therefore a definition of population is needed that will minimize arbitrariness. Davis and Nixon[11] define populations as "arenas in which sexual reproduction, genetic recombination and character fixation occur." Once the individuals have been designated to populations, attributes are scored and their frequencies are recorded within the population. Each attribute is then scored in each population as present, absent or variant in the population. Populations are then compared to each other to determine if they are distinguishable on the basis of at least one attribute. The diagnostic attribute has to be absolute, that is, the original state or a transformed state must be fixed (i.e. present in all individuals) in one population and absolutely absent in the other. When fixed differences do not appear, the populations are considered a single unit and a similar profile for these single unit multipopulations is generated. The procedure continues until no further aggregation is warranted.

The errors associated with this technique are twofold.[11] Incorrect homology assessment will adversely affect this analysis (as well as any phylogenetic analysis). For instance, many times with isozymes it is difficult to assess the relatedness of transformed character states. If a population was polymorphic for allele A and its transformed state, allele B, and one could not determine that the transformed state was allele B then this locus would be variant, designated as a trait and would not be recognized as a potential character for this population. Another source of error with this technique is undersampling. Sampling of too few individuals will have either of two effects. The lack of detection of rare alleles will result in the recognition of too many differentiated populations and hence inflate the number of recognized characters. Undersampling of individuals could result in the recognition of too few aggregations in cases where an attribute is rare in some populations and common in others, in which case these populations could be taken to be part of different aggregations. Sampling of too few populations will result in an inflated number of characters if populations with intermediate composition of attributes are not included in the analysis. Finally, if too few attributes are sampled, the probability of detecting diagnostic characters is low and the differentiation of populations will go undetected.

Individuals as Terminals

The use of individuals as terminals has been a common practice in modern molecular systematics,[1,3,42] but the theoretical reasoning behind this practice has been unclear. The major question with using individuals as terminals is then: Are the results of these cladistic analyses congruent with the descent hierarchy (phylogenetic analysis)? Vrana and Wheeler[41] discuss the theoretical implications of using individuals as terminals in cladistic analysis. Following Nelson's[28] view of monophyly they state that "taxa are seen as relationships of organisms, rather than groups of them" (Vrana and Wheeler, 1992; p 67). Nelson[28] and Vrana and Wheeler[41] make the counterintuitive philosophical suggestion that in phylogenetic analysis characters do not belong to taxa, rather taxa belong to characters. They further argue that the traditional "line of death" should not demarcate a level below which systematics cannot be done. Vrana and Wheeler[41] refer to Hennig's[19] Figure 6 and claim that it is heavily assumption laden. One assumption of the model is that some "inexplicable" process produces a clean break between tokogeny and phylogeny or between reticulation and divergence. Other assumptions are that the break is obvious and easily recognizable and that there is a constancy of process across all sexually reproducing biparental organisms.[41] Due to the assumptions inherent in the model in Figure 6 of Hennig[19] they claim that such a model is an unfounded statement in need of verification.[28] In other words the "level at which reticulation occurs and pattern is not discernible can only be determined empirically."[41] They further claim that there is no way of knowing before an analysis where the level of reticulation occurs and hence no way of *a prior* knowing where to apply the "line of death." Cladistic analysis in the scenarios of Nelson[28] and Vrana and Wheeler[41] will imply genealogy and can be accomplished regardless of the rank given to a taxon.

Coalescent Theory

Coalescent theory was first developed by Kingman[21,22] and later expanded on by Hudson,[20] Ewens,[17] Crandall and Templeton.[9] The basic goal of coalescent theory is to describe genealogy back through the time of the coalescent. The determination of the age of the coalescent is a critical component of this theory.[2,9,20] Assessing the depth of a coalescent event (i.e. the age of the event) for two terminals will indicate the degree of relatedness of the terminals.

Crandall and Templeton[9] suggest that coalescent theory can address several questions about the use of frequency information in reticulate systems. For instance, haplotype frequency is used to refine cladistic position in a network of haplotypes and frequency is also used to relate the number of character state changes in these networks. Several assumptions are made in the coalescent model and in this framework the raw materials of such an analysis are still allele frequencies. It appears to us that the criticisms leveled at the use of allele frequencies in phylogeny reconstruction outlined above are applicable for this application of coalescent theory too. However, Crandall and Templeton[9] have shown using empirical data that coalescent theory has some readily testable predictions.

Coalescent theory may have its greatest application at refining networks of haplotypes composed of individuals used as terminals[9,36,37] and at establishing hypotheses about coalescent events. These coalescent events then might be defined as indicators of speciation events under the assumptions of the "genealogical species concept."[2] Concordance of gene genealogies is an important aspect of this approach. In fact, Baum and Shaw propose the use of multiple gene phylogenies in coalescent analysis followed by consensus techniques to assess concordance of the gene systems. The point of coalescence for multiple gene phylogenies is taken as the divergence point of new phylogenetic lineages.

Illustration of the Methods: Beetles and Flies

1) The Northeastern Beach Tiger Beetle (*Cicindela dorsalis*): We have described the population biology and genetics of *C. dorsalis* on the Atlantic Coast and Gulf of Mexico of North America.[38,39,40] *C. dorsalis* is a polytypic species complex with a wide geographic range from Massachusetts to Veracruz, Mexico (Figure 1, Table 1). Our analyses are based on an extensive survey of mtDNA covering individuals from representative populations over the entire range of *C. dorsalis*. In a combination of sequencing and RFLP techniques 21 different haplotypes in approximately 400 individuals could be distinguished.

2) Hawaiian *Drosophila*: mtDNA RFLP data[13,14] from two species of Hawaiian *Drosophila* are also examined in detail. They are *D. silvestris* and *D. heteroneura*. For a review of the biology and genetics of these two species see Carson.[5,7] These two species are chromosomally homosequential[6] and have overlapping isozyme profiles. In a population aggregation analysis[11] the allozyme and chromosomal inversion attributes are all determined to be "traits." We examined eight populations from various well separated localities on the Big Island of Hawaii

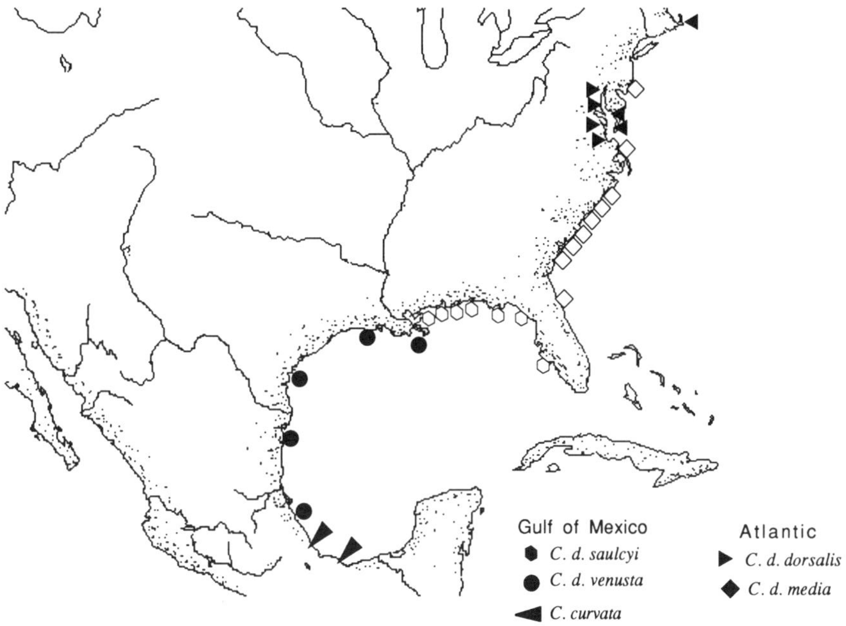

Figure 1. Map of the North America showing the geographic distribution of the *C. dorsalis* populations.

(Figure 2, Table 2). Twenty four restriction enzymes were used in the analysis and all sites were mapped by double digestion approaches.

The Analyses

There are two possible ways of presenting these data. Each nucleotide site or RFLP site could be treated as a character. The other method of treatment concerns the designation of haplotypes as characters. We present the analyses using each individual nucleotide and restriction site as characters in all of the following analyses.

C. dorsalis *on the East Coast of the United States*

Table 1 lists the populations and the number of individuals examined from each population of *C. dorsalis*. At the Atlantic coast of North America there are a total of 15 haplotypes from 24 populations from over 350 individuals. An additional five haplotypes were discovered in a less comprehensive survey of 45 individuals in populations from the Gulf of Mexico. Figure 3 shows a parsimony analysis of the DNA sequences using individuals as terminals. There is resolution in this cladogram at only eight nodes, the most conspicuous being the node that

Table 1. Populations, numbers of individuals, location (latitude/longitude), and haplotypes in the C. dorsalis *study. The population codes refer to populations in all subsequent figures*

Population	Pop. Code	Lati/Longi		Haplotypes	*n*
New England (*C.d.dorsalis*)					
Martha's Vineyard, MA	MV	41.3	70.7	A0,A1,A2	22
Chesapeake Bay, west (*C.d.dorsalis*)					
Western Shores, MD	WS	38.4	76.4	A3	4
Flag Pond, MD	FP	38.4	76.4	A3	20
Smith Point, VA	SP	37.9	76.4	A3,A5	25
Fleeton, VA	FL	37.8	76.4	A3,A4	10
Kilmarnock, VA	KI	37.7	76.4	A3,A4,A5	26
Winter Harbour, VA	WH	37.4	76.4	A3,A4,A5	24
Grandview Park, VA	GVP	36.9	76.3	A5,A7,A8	21
Chesapeake Bay, east (*C.d.dorsalis*)					
Silver Beach, VA	SB	37.5	75.9	A3,A5	21
Picketts Harbor, VA	PH	37.2	76.0	A3,A3a,A5,A7,A8	27
Wise Point, VA	WP	37.1	76.0	A3,A3a,A5,A7,A8	26
Atlantic Coast (*C.d.media*)					
Little Egg Island, NJ	LEI	39.4	74.3	A5	1
Chincoteague, VA	CH	37.9	75.4	A5	21
Fisherman Island, VA	FI	37.1	76.0	A5,A6	29
Fort Story, VA	FS	36.9	76.0	A5,A7	3
Long Beach, NC	LB	33.8	78.3	A7,A8,A10	23
Holden Beach, NC	HNB	33.8	78.4	A7,A8,A7a,A10	16
Huntington State Park, SC	HSP	33.5	79.1	A5,A7,A8,A7a,A9, A10,A11,A13	26
Pawleys Island, SC	PI	33.4	79.2	A7,A8,A7a,A10	26
Folly Beach, SC	FB	32.6	79.9	A10	3
Edisto Beach, SC	EB	32.5	80.3	A10	4
Tybee Island, GA	TI	31.2	81.3	A10,A11	18
Little Talbot Island, FL	LTI	30.4	81.3	A10,A11,A12	22
Cape Canaveral, FL	CC	28.8	80.7	A12	1
Gulf of Mexico (*C.d.saulcyi, C.d.venusta*)				G0,G1,G2,G3,G4,G5	45

separates the haplotypes encountered in the Gulf of Mexico and the Atlantic Coast populations. Within either of these main clades there is little resolution in this type of analysis. Most of the haplotypes are not restricted to individual populations or geographic regions. Thus, parsimony analysis on individual sequences subdivides the mtDNA haplotypes but these subdivisions do not correspond to monophyletic groups that are congruent with our population designations.

For the qualitative Hennigian analysis, the frequency analysis and the population aggregation analysis, individuals were designated as members of populations based on geographic locality and isolation of beach habitats (24 populations of

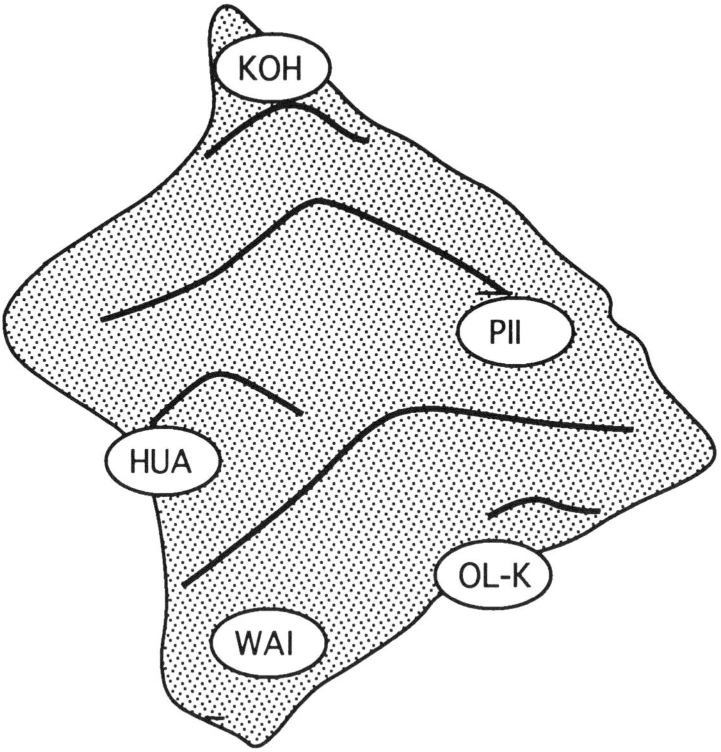

Figure 2. Map of the Big Island of Hawaii showing the location of the populations used in this part of the study. Abbreviations are as in Table 2.

the Atlantic lineage in total). The presence or absence of individual nucleotide positions were scored. The treatment of nucleotides as characters for qualitative Hennigian analysis essentially results in a mostly unresolved cladogram and does not group any of the populations into monophyletic clades (Figure 4).

Swofford and Berlocher's[35] FREQPARS program was used to construct maximum parsimony trees from allele frequency data. Each nucleotide site was treated as a locus and the presence or absence of a G, A , T, C was assessed and frequencies for these "alleles" were calculated. The tree generated by this program as shown in Figure 4, demonstrates the ability of the FREQPARS procedure to extract more resolution from the mtDNA data set. This tree is highly concordant with the geographical locality of these populations. In particular, the more ancestral populations in the cladogram are Southeastern Atlantic Coast populations while the more derived populations are found in the northernmost reaches of the distribution of this species.

Population aggregation analysis was performed on this data set. Figure 5 shows the patterns obtained using this technique. The results of this analysis were

Table 2. Populations, numbers of individuals and location in the Hawaiian Drosophila study. Locations refer to those parts of the Big Island of Hawaii in Figure 2, unless noted otherwise.

Species	Population Abbreviation	Location	*n*
2 bristle rows on foreleg			
silvestris	OL-K	Olaa/Kilauea	9
silvestris	PII	Piihonua	10
silvestris	KH	Kohala	2
3 bristle rows on foreleg			
silvestris	HUA	Hualalai	10
2 bristle rows on foreleg			
heteroneura	OL	Olaa	1
heteroneura	HUA	Hualalai–1	2
heteroneura	HUA	Hualalai–2	2
heteroneura	WAI	Waihaka	13
2 bristle rows on foreleg			
planitibia	MAUI	Waikamoi, Maui	4

straightforward and reduced the number of informative nucleotide sites from 25 to 1 character. There are an additional 39 characters that distinguished the Atlantic haplotypes from the Gulf of Mexico lineage. The most striking result from this analysis of populations in the Atlantic is that the Martha's Vineyard population is the only population that is diagnosable, and this is based on the presence of only a single nucleotide character. On the basis of this analysis we inferred that the Martha's Vineyard population which has attracted the attention of conservation biologists is a diagnosable and unique evolutionary unit.[40]

Hawaiian Drosophila

Table 2 lists the populations and numbers of individuals that were analyzed for this study. Characters and character states are from mtDNA RFLP data and a discussion of the data has appeared in previous publications.[13,14] In those papers the analyses were primarily phenetic. In addition, another analysis of these two species was accomplished using more character state information by DNA sequencing techniques but for fewer individuals.[16] All of the analyses for this part of the study were accomplished using each restriction site as a character. There were over 120 mapped restriction sites for these populations of which 39 were phylogenetically informative over all individuals. *D. planitibia* was used as an outgroup. This species resides on Maui and is one of two plausible outgroup taxa that could be used.

Figure 6 shows the analysis of the data set using individuals as terminals. Due to the large number of individuals, heuristic searches with multiple repetitions

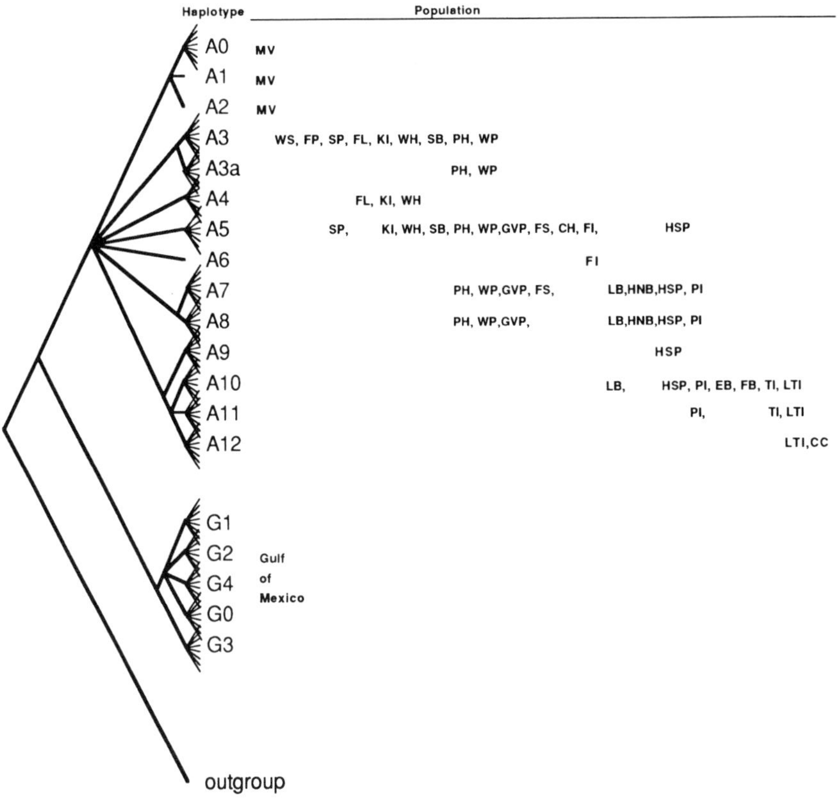

Figure 3. Consensus cladogram generated using individuals as terminals for the *C. dorsalis* study. Haplotypes are described in the text. The branches emanating from the tips of the tree refer to the presence of multiple individuals with those haplotypes. Abbreviations for the populations are given in Table 1.

of random addition of taxa were used to search the available tree space. Over 800 parsimony trees were found for these data. Figure 6 shows the strict consensus tree for these parsimony trees. There are three major clades of flies seen in this individuals as terminals tree. The first corresponds to *D. silvestris* from the Hilo side of the Big Island, the second corresponds to the *D. heteroneura* and the third corresponds to *D. silvestris* from the Kona side of the island. The relationships of the three clades to each other is however unresolved as evidenced by the trichotomous branching of the clades.

Eight populations were used in the qualitative Hennigian analysis (Figure 7). This figure demonstrates the existence of three clades once again and suggests that there is no resolution as to the relatedness of the three clades to each other, a result very similar to the individuals as terminals analysis. The allele frequency analysis yielded a tree (Figure 7) which also shows the three major clades. This

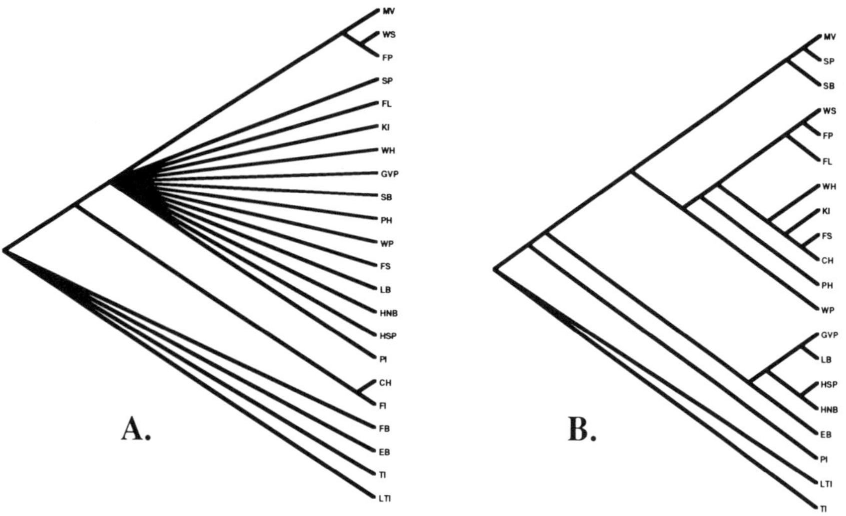

Figure 4. Panel A: The qualitative Hennigian analysis tree for the *C. dorsalis* study; Panel B: the FREQPARS tree generated from frequency data for the *C. dorsalis* study. Due to the fact that the FREQPARS program can handle only twenty taxa we arbitrarily removed two closely related taxa from the analysis in B. These two taxa were FI and FB. Abbreviations are as in Table 1.

cladogram, however, shows a sister group relationship for *heteroneura* with the Kona side *silvestris*. Although the *heteroneura* are morphologically very different from the *silvestris* in head shape and body coloration, the hypothesis in Figure 7 is interesting and plausible, because the *silvestris* on the Hilo side have a derived morphological trait of three rows of bristles on the forelegs of males. Kona side males as well as *heteroneura* and the outgroups have only two rows of bristles on the foreleg.[6]

Population aggregation analysis detected seven characters (Figure 8) that show fixation in all populations examined. This analysis aggregates the three Hilo side populations as a unit on the basis of two of the characters (Table 3). The Waihaka *heteroneura* population is unique at three characters and can hence be diagnosed as a separate unit. One character also aggregates the *silvestris* from Hualalai with the other *silvestris* and one character aggregates this *silvestris* population with *heteroneura*. Figure 7 shows a cladogram constructed from these seven characters. The cladogram shows clearly the aggregation of the three Hilo side *silvestris* populations, while the other populations are basal and unresolved. This result, although not well resolved, is not at odds with the hypotheses generated by the other methods.

It should be noted that the above analyses all use mtDNA character sets, and should be considered as indicators of how asexual systems will behave when inferring the phylogenetic and cladistic relationships present in the character set.

Pop	Ind	Attribute																									Characters
		1	2	3	4	5	6	7	8	9	10	11	12	13	14	15	16	17	18	19	20	21	22	23	24	25	
MV	22	±	**0**	1	1	1	±	1	1	0	1	1	1	1	1	1	1	1	1	1	1	1	1	1	1	0	1
WS	4	1	**1**	1	1	1	1	1	1	1	1	1	1	1	0	1	1	0	1	1	1	1	1	1	1	0	0
FP	20	1	**1**	1	1	1	1	1	1	1	1	1	1	1	0	1	1	0	1	1	1	1	1	1	1	0	0
SP	25	1	**1**	1	1	1	1	1	1	1	1	1	±	1	±	1	1	±	1	1	1	±	1	1	1	±	0
FL	10	1	**1**	1	1	1	1	1	1	1	1	±	±	1	±	1	1	±	1	1	1	±	1	1	1	±	0
KI	26	1	**1**	1	1	1	1	1	1	1	1	±	±	1	±	1	1	±	1	1	1	±	1	1	1	±	0
WH	24	1	**1**	1	1	1	1	1	1	1	1	±	±	1	±	1	1	±	1	1	1	±	1	1	1	±	0
GVP	21	1	**1**	1	1	1	1	1	±	1	1	±	±	1	1	1	1	1	1	±	1	±	1	±	1	1	0
SB	21	1	**1**	1	1	1	1	1	1	1	1	1	±	1	±	1	1	1	1	1	1	±	1	1	1	±	0
PH	27	1	**1**	1	1	1	1	1	±	1	1	±	±	1	±	1	1	1	1	1	±	±	1	±	1	±	0
WP	26	1	**1**	1	1	1	1	1	±	1	1	±	±	1	±	1	1	1	1	±	±	±	1	±	1	±	0
CH	21	1	**1**	1	1	1	1	1	1	1	1	1	0	1	1	1	1	1	1	1	1	0	1	1	1	1	0
FI	29	1	**1**	1	1	1	1	1	1	1	1	1	0	1	1	±	1	1	1	1	1	0	1	1	1	1	0
FS	3	1	**1**	1	1	1	1	1	±	1	1	±	±	1	1	1	1	1	1	±	1	±	1	1	1	1	0
LB	23	1	**1**	1	±	±	1	±	±	±	±	±	1	1	1	1	±	1	±	±	1	±	1	±	1	1	0
HNB	16	1	**1**	1	±	±	1	±	±	±	±	±	1	1	1	1	±	1	±	±	1	±	1	±	1	1	0
HSP	27	1	**1**	±	±	±	1	±	±	±	±	±	±	1	1	1	±	1	±	±	1	±	±	±	±	1	0
PI	26	1	**1**	1	±	±	1	±	±	±	±	±	1	1	1	1	±	1	±	±	1	±	1	±	1	1	0
FB	3	1	**1**	1	0	0	1	0	1	0	0	1	1	1	1	1	0	1	0	1	1	0	1	0	1	1	0
EB	4	1	**1**	1	0	0	1	0	1	0	0	1	1	1	1	1	0	1	0	1	1	0	1	0	1	1	0
TI	18	1	**1**	±	0	0	1	0	1	0	0	±	1	1	1	1	0	1	0	1	1	0	1	0	1	1	0
LTI	22	1	**1**	1	0	0	1	0	1	0	0	±	1	1	1	1	0	1	0	1	1	0	1	0	1	1	0
Gulf of Mexico																											39

Figure 5. Population aggregation analysis of the *C. dorsalis* data set. The numbers across the top (1–25) indicate the variable nucleotide sites in the study. A + indicates the presence of the attribute in the populations listed in the far left column while a 0 indicates its absence and a +/− indicates that the population is polymorphic for that attribute. The number of individuals from each population is listed in the "Ind" column on the left. The far right column (characters) indicates the number of attributes that can be considered characters for the global data set and define the population as diagnosable.

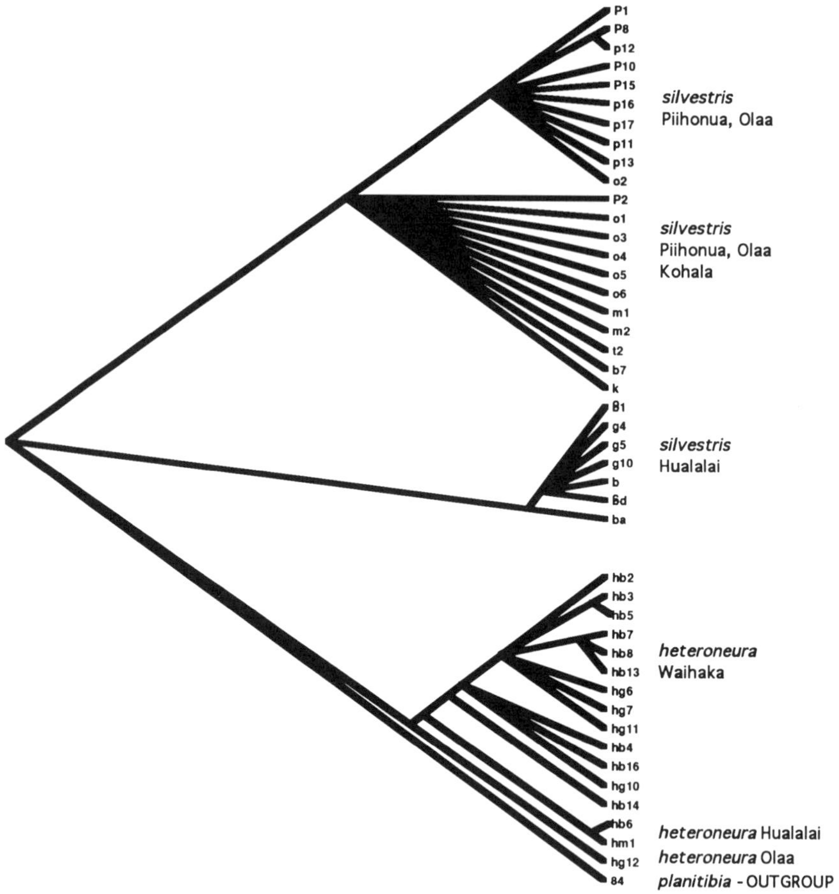

Figure 6. Consensus cladogram generated using individuals as terminals for the Hawaiian *Drosophila* data set. 865 equal length trees were generated from the data set using the Heuristic search option in PAUP and the random taxa addition method.

Davis and Nixon[11] discuss in detail the pitfalls of using a maternally inherited gene in phylogenetic analysis in the Hennigian context, and the gene tree species tree problem in phylogenetic analysis has been discussed at length.[20]

Summary

Certain assumptions and philosophical stands have to be taken when considering which methodology to use. The qualitative Hennigian analysis, FREQPARS and population aggregation analysis methods all assume that the investigator knows what a "population" is. The definition of population given by Davis and Nixon[11]

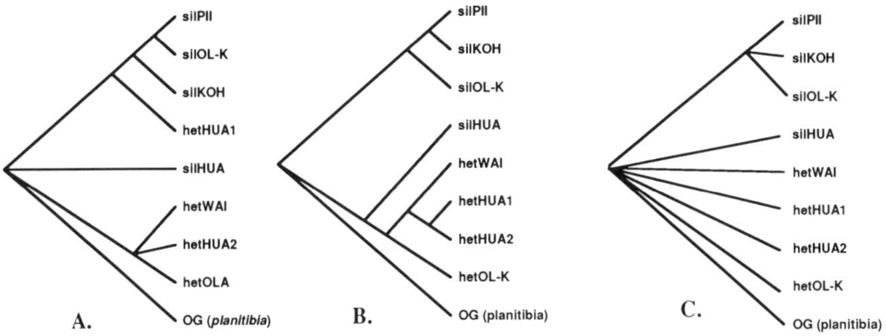

Figure 7. Panel A: Qualitative Hennigian analysis of the RFLP data for the Hawaiian *Drosophila* data set. Panel B: FREQPARS for the Hawaiian *Drosophila* data set; Panel C: Consensus cladogram generated using the population aggregation analysis characters from Table 3.

is objective and an excellent working definition. This definition could however be criticized as operationally difficult to implement. The more common use of population concerns an operationally useful designation of population using geographic demarcation as in the two examples discussed in this paper. The common operationally useful definition and the non-arbitrary definition in Davis and Nixon[11] often times are not the same. Coalescent theory as used by Baum and Shaw[2] and using individuals as terminals, on the other hand, make no assumptions of what a population is. Individuals and their relationships to each other are used to test hypotheses of hierarchy in the individuals as terminals approach. Coalescence of individuals to a common branching point in the coalescent process is an indication of common ancestry for groups of individuals.

In both the *Cicindela* and *Drosophila* examples, using individuals as terminals resulted in cladograms with several resolved nodes, but concordance with strict population demarcations were observed only in a few instances. Vrana and Wheeler and Nelson would take these results as indication of a lack of discrete population level structure. In other words, the structure inherent at the individual level does not translate into population level structure. Hypotheses concerning the existence of discrete populations in both of these systems would be rejected for all population designations except the Martha's Vineyard in the beetle example and the Hualalai *silvestris* population.

Two techniques (qualitative Hennigian analysis and frequency parsimony) attempt to deal with polymorphism using relatively different approaches. Both of these techniques result in highly resolved cladograms that are concordant with what one might expect for the relationships of individuals in both the beetle and Hawaiian *Drosophila* populations. The results for the beetle populations are perhaps more interesting because of the north to south divergence of the populations inferred from these analyses.

Population aggregation analysis results in the elimination or loss of resolution

```
D. silvestris - Piihonua
                    *  **    *      **   ****   *   *
P1              101000100000010100111101001010011010001
P2              101000111000010100110001010100010010001
P8              101010101000010100110001011010010010001
P10             101000101000000100111101010010010010101
P15             101000111000010100111101010010010010001
p16             101000111000010100111101001110010010001
p17             101000111000010100111101010010010010001
p11             101000101000010100111101010010011010001
p12             101010101000010100111101011010010010001
p13             101000111000010100111101010010011010001

D. silvestris - Olaa
                   *         * * * *     ****   *   *
o1              101000100000000100111001010100010010101
o2              101000101000010100101001001010011010001
o3              101000101000000000111100101110001001010 1
o4              101000101000010100111001000100010010001
o5              101100101000010100101001010100010010001
o6              101000101000010100111001010100011010001
m1              101000101000010000111001010100010010001
m2              101100101000000000111100101010001001000 1
t2              101000101000010100111001111100011010001

D. silvestris - Kohala
                  *  *       *            *     *
b7              101000101000000100111001011100010010001
ko              100001101000010100111001010100011010001

D. silvestris - Hualalai
                    **   *  * *  *  * * *    ** *  ****
b1              101010101101000000111101111011010000011 0
b2              101000101001001000111001010111001001011
g4              101000101101000000111101111011010000011 1
g5              101000101101000000111001111011010000011 1
g10             101000101101000000111101111011010000110 1
ba              101000101101001000111101111011010000001 1
bc              101000101101000000111101111011010000011 1
bd              101000101101000000111101111011010000011 1
b9              101001101001100000110011010111000000110

D. heteroneura - Waihaka
                **        *         *            * *  *
hb2             100000001001010010010000010111000001110
hb3             100000001001010010011000010111000001110
hb4             100000001001010010010000010111000001111
hb5             100000001001010010011000010111000101110
hb7             000000001001010010010000010111000001110
hb8             000000001001010010010000010111000001110
hb13            000000001001010010010000010111000001110
hb14            110000001011010010010000010111000001111
hb16            100000001001010010010000010111000001111
hg6             100000001001010010010000010111000001110
hg7             100000001001010010010000010111000000110
hg10            100000001001010010010000010111000101111
hg11            100000001001010010010000010111000001110

D. heteronuera - Hualalai1
                                    **
hb6             100000001011001000110001010111001001001
hm1             100000001011001000110001010111001001111

D. heteroneura - Hualalai2
                                   * *
hm2             100001001000010100110001010110101000111
hg3             100001001000010100110001010110101001101

D. heteroneura - Olaa

hg12            110000001011100001111001010111000000111

D. planitibia - Maui
84              101011000001010001101001010000101000011
```

Figure 8. Population aggregation analysis of the Hawaiian *Drosophila* data set. Thirty-nine phylogenetically informative sites are shown for each of the individuals in the study. A zero indicates the absence of a restriction site and a one indicates the presence. Stars above the sites for each population indicate those sites that are variant within each of the designated populations.

Table 3. "Characters" from population aggregation analysis. The seven sites noted in Table 3 as invariant for a population are tabulated in this table. A zero indicates the absence of a character and 1 indicates the presence of the character.

Species	Locality	Characters
silvestris	Piihonua	1 0 0 1 1 1 1
silvestris	Olaa	1 0 0 1 1 1 1
silvestris	Kohala	1 0 0 1 1 1 1
silvestris	Hualalai	1 1 0 1 1 0 0
heteroneura	Waihaka	0 1 1 0 0 0 0
heteroneura	Hualalai-1	0 1 0 1 1 0 0
heteroneura	Hualalai-2	0 0 0 1 1 0 0
heteroneura	Olaa	0 1 0 1 1 0 0
planitibia	Maui	0 1 0 1 1 0 0

(Figures 5 and 8 and Kluge[23]) in both of these examples. If we accept a difference between phylogenetic analysis and cladistic analysis and the two conditions for Hennigian phylogenetic analysis outlined by Davis and Nixon[11] we are forced to accept these results. The general lack of resolution in the results using this approach can be taken as an indicator of the level of reticulation or tokogeny that exists in these populations. It is interesting that despite the propensity for loss of resolution using this approach, certain populations do aggregate in both the *Drosophila* example and the *Cicindela* example.

A comparison of the boundaries between tokogeny and phylogeny in both of these examples reveals some interesting patterns. In the tiger beetle example, reticulation occurs among the Atlantic coast populations that results in the overlapping cladogram for individuals as terminals (Figure 3) and in the inability of the population aggregation analysis to detect characters (Figure 5). In the *Drosophila silvestris* populations, reticulation is also evident but only on one side of the island of Hawaii (Figure 6). More characters were detected in this example (Figure 8) than in the tiger beetle example indicating less reticulation. It is possible that the *Drosophila* populations are at a more advanced stage of differentiation than the tiger beetle populations or that undersampling of the *Drosophila* populations produces this result.

Coalescent techniques were not attempted for these examples because of the lack of multiple gene genealogies for these organisms. However as more information accumulates from nuclear genes for these and other organisms, reticulating gene systems will be of major interest in phylogenetic analysis at the individual and population level. According to several authors, the coalescent method offers a more detailed and precise methodology for understanding the phylogenetic history of individuals within populations and between populations. Baum and Shaw[2] suggest a way of reconciling competing gene genealogies that is consistent with coalescent theory and with Hennigian analysis. They describe a theoretical system where DNA sequences are obtained for four nuclear genes from several individuals. Each gene tree is constructed individually and a consensus tree is inferred to describe the coalescent. The coalescent is implied by the presence of an unresolved polytomy of several individuals.

Consideration of the modern systematic analysis and the theory behind it is necessary before proceeding to conclusions about the phylogenetic history of individuals within populations and of populations. If a gene genealogy is desired then the strictures of cladistic analysis and the use of individuals as terminals accommodate this nicely. If the phylogenetic history of a group of individuals or of populations is desired, then the phylogenetic systematic framework described by Davis and Nixon[11] must be considered.

Acknowledgments

We would like to thank John Gatesy, Darrel Frost, Paul Vrana and Charlie Wray for comments on early drafts of the manuscript. We would also like to thank Brian Golding and the CIAR for their invitation to this symposium.

References

1. J. C. Avise, J. Arnold, R. M. Ball, E. Bermingham, T. Lamb, J. E. Neigel, C. A. Reeb, and N. C. Saunders. (1987). Intraspecific phylogeography: The mitochondrial bridge between population genetics and systematics. *Annual Review of Ecology and Systematics* 18, 489–522.
2. D. A. Baum, and K. L. Shaw. (1994). Genealogical perspective on the species problem. Experimental and Molecular Approaches to Plant Biosystematics. ed. by P.C Hoach , A.G. Stevenson and B.A. Schaal. St. Louis Monographs in Systematics, Missouri Botanical Garden (in press).
3. R. L. Cann, M. Stoneking, and A. C. Wilson. (1987). Mitochondrial DNA and human evolution. *Nature* 325, 31–36.
4. J. M. Carpenter, J. E. Strassmann, S. Turillazzi, C. R. Hughes, C. R. Solis, and R. Cervo. (1993). Phylogenetic relationships among paper wasp social parasites and their hosts (Hymenoptera: Vespidae; Polistinae). *Cladistics* 9, 129–146.
5. H. L. Carson. (1982). Evolution of Drosophila on the newer Hawaiian volcanoes. *Heredity* 48, 3–25.
6. H. L. Carson. (1987). Chromosomal evolution of Hawaiian Drosophila. *Trends in Ecology and Evolution* 2, 200–206.
7. H. L. Carson. (1990). Evolutionary process as studied in population genetics: clues from phylogeny. pp. 129–156 in Oxford Surveys in Evolutionary Biology, Volume 7 ed. by J. Antonovics and D. Futuyma. Oxford University Press, New York.
8. J. Cracraft. (1983). Species concept and speciation analysis. *Current Ornithology* 1, 159–187.
9. K. A. Crandall, and A. R. Templeton. (1993). Empirical tests of some predictions

from coalescent theory with applications to intraspecific phylogeny reconstruction. *Genetics* 134, 959–969.

10. B. I. Crother. (1990). Is "some better than none" or do allele frequencies contain phylogenetically useful information? *Cladistics* 6, 277–281.
11. J. I. Davis, and K. C. Nixon. (1992). Populations, genetic variation, and the delimitation of phylogenetic species. *Systematic Biology* 41, 421–435.
12. K. de Queiroz, and M. J. Donoghue. (1990). Phylogenetic systematics and the phylogenetic species concept. *Cladistics* 4, 317–338.
13. R. DeSalle, and L. V. Giddings. (1986). Discordance of nuclear and mitochondrial DNA phylogenies in Hawaiian *Drosophila. Proceedings of the National Academy of Science* 83, 6902–6906.
14. R. DeSalle, L. V. Giddings, and K. Y. Kaneshiro. (1986). Mitochondrial DNA variability in natural populations of Hawaiian *Drosophila*. II. Genetic and phylogenetic relationships of natural populations of *D. silvestris* and *D. heteroneura. Heredity* 56, 87–96.
15. R. DeSalle, and A.R. Templeton. (1988). Founder effects and the rate of mitochondrial DNA evolution in the Hawaiian Drosophila. *Evolution* 42, 1076–1084.
16. R. DeSalle, and A. R. Templeton. (1992). The mtDNA genealogy of closely related *Drosophila silvestris. Journal of Heredity* 83, 211–216.
17. W. J. Ewens. (1990). Population genetics theory-the past and the future, pp 177–227 in *Mathematical and Statistical Developments of Evolutionary Theory,* edited by S. Lessard. Kluward Academic Publishers, New York.
18. J. S. Farris. (1983). The logical basis of phylogenetic inference. In: (N. I. Platnick and V. A. Funk, ed.) *Advances in Cladistics*. Volume 2. Columbia University Press, New York.
19. W. Hennig. (1966). Phylogenetic Systematics. University of Illinois Press, Urbana.
20. R. R. Hudson. (1990). Gene genealogies and the coalescent process. *Oxford Survey in Evolutionary Biology* 7, 1–44.
21. J. F. C. Kingman. (1982a). The coalescent. *Stochast. Proc. Appl.* 13, 235–248.
22. J. F. C. Kingman. (1982b). On the genealogy of large populations. *J. Appl. prob.* 19A, 27–43.
23. A. G. Kluge. (1989). Metacladistics. *Cladistics* 5, 291–294.
24. W. P. Maddison, and D. R. Maddison (1992) MacClade version 3.01. Sinauer Associates, Sunderland, Massachusetts.
25. M. Mickevich. (1982). Transformation series analysis. *Systematic Zoology* 31, 461–478.
26. M. F. Mickevich, and M. S. Johnson. (1976). Congruence between morphological and allozyme data in evolutionary inference and character evolution. *Systematic Zoology* 25, 260–270.
27. M. Nei. (1987). Molecular evolutionary genetics. Columbia University Press, New York.

28. G. Nelson. (1989). Cladistics and evolutionary models. *Cladistics* 5, 275–289.

29. G. J. Nelson, and N. I. Platnick. (1981). Systematics and biogeography: Cladistics and vicariance. Columbia University Press, New York.

30. K. C. Nixon, and J. I. Davis. (1991). Polymorphic taxa, missing values and cladistic analysis. *Cladistics* 7, 233–241.

31. K. C. Nixon, and Q. D. Wheeler. (1990). An amplification of the phylogenetic species concept. *Cladistics* 6, 212–223.

32. J. C. Patton, and J. C. Avise. (1983). An empirical evaluation of qualitative Hennigian analyses of protein electrophoretic data. *Journal of Molecular Evolution* 19, 244–254.

33. J. C. Patton, R. J. Baker, and J. C. Avise. (1981). Phenetic and cladistic analyses of biochemical evolution in peromyscine rodents. In: (M. H. Smith and J. Joule, ed.) *Mammalian population genetics*. University of Georgia Press, Athens.

34. D. L. Swofford, Phylogenetic Analysis Using Parsimony (PAUP); program and documentation. Natural History Survey, University of Illinois, Champaign, Illinois.

35. D. L. Swofford, and S. H. Berlocher. (1987). Inferring evolutionary trees from gene frequency data under the principle of maximum parsimony. *Systematic Zoology* 36, 293–325.

36. A. R. Templeton, E. Boerwinkle, and C. F. Sing. (1987). A cladistic analysis of phenotypic associations with haplotypes inferred from restriction endonuclease mapping and sequencing data. I. Basic theory and an analysis of alcohol dehydrogenase activity in *Drosophila*. *Genetics* 117, 343–351.

37. A. R. Templeton, K. A. Crandall, and C. F. Sing. (1992). A cladistic analysis of phenotypic associations with haplotypes inferred from restriction endonuclease mapping and sequencing data. III. Cladogram estimation. *Genetics* 132, 619–633.

38. A. P. Vogler, and R. DeSalle. (1993a). Mitochondrial DNA evolution and the application of the phylogenetic species concept in the *Cicindela dorsalis* complex (Coleoptera: Cicindelidae). In: (K. Desender, ed.) *Carabid beetles: ecology and evolution*. Kluwer Academic Press, Dordrecht, The Netherlands.

39. A. P. Vogler, and R. DeSalle. (1993b). Phylogeographic patterns in coastal North American Tiger Beetles, *Cicindela dorsalis* inferred from mitochondrial DNA sequences. *Evolution* 47, 1192–1202.

40. A. P. Vogler, R. DeSalle, T. Assmann, C. B. Knisley, and T. D. Schultz. (1993). Molecular population genetics of the endangered tiger beetle, *Cicindela dorsalis* (Coleoptera: Cicindelidae). *Annals of the Entomological Society of America* 86, 142–152.

41. P. Vrana, and W. C. Wheeler (1992) Individual organisms as terminal entities: Laying the species problem to rest. *Cladistics* 8, 67–72.

42. A. C. Wilson, R. L. Cann, S. M. Carr, M. George, U. B. Gyllenstein, K. M. Helm-Bychowski, R. G. Higuchi, S. R. Palumbi, E. M. Prager, R. D. Sage, and M. Stoneking (1985) Mitochondrial DNA and two perspectives on evolutionary genetics. *Biological Journal of the Linnaean Society* 26, 375–400.

14

The Divergence of Halophilic Superoxide Dismutase Gene Sequences: Molecular Adaptation to High Salt Environments

Patrick P. Dennis

During the divergence of homologous sequences, nucleotide substitutions can become fixed either through random processes or by virtue of positive selection.[6,9,14] The underlying components that together contribute to a positive selection value for a given substitution can be many and varied. They could include selection (i) for altered structure or function within the encoded protein, (ii) for codon utilization as it relates to either efficiency or accuracy of mRNA translation, and (iii) for regulatory signals or secondary structure embedded in the DNA or mRNA sequence.

Because of past selection and because physical environments tend to change slowly, the fitness of contemporary DNA coding sequences are already near optimal. Most mutations that occur produce suboptimal variants and are rapidly eliminated by negative selection. A few mutations are essentially neutral and may be either fixed or eliminated by random processes. Most often, the mutations that are fixed represent third position synonymous substitutions in codons that have negligible impact on translation efficiency and accuracy; less often, they are nonsynonymous substitutions and result in an amino acid replacement at a noncritical position in the protein. On rare occasions, a mutation that significantly improves fitness can occur. It is usually a nonsynonymous substitution that replaces a less favorable amino acid in the protein with a more favorable one. Consequently, homologous genes initially diverge primarily through the accumulation of third position synonymous substitutions, and the similarity between their DNA sequences (measured as percent nucleotide identity) degenerates more rapidly than the similarity between their encoded proteins (measured as percent amino acid identity). Random processes as well as changing selective processes over a long period of time eventually result in the accumulation of a significant number of nonsynonymous substitutions. Only then do the protein sequences become less similar than the encoding DNA sequences.

There is ample evidence to support this general picture. For both mammals and eubacteria, the rate of substitution at synonymous positions within most protein encoding genes has been estimated to be about 0.8% per million years of divergence. In contrast, the rate of nonsynonymous substitutions is in general estimated to be between five to twentyfold lower.[14] For most homologous sequences present in the data bases, the crossover point at which protein sequence becomes less similar than the encoding DNA sequence occurs when about 35% of the nucleotide positions contain substitutions and are no longer identical (R. Doolittle, personal communication of an unpublished result).

Within this general background, I wish to review the divergence of seven superoxide dismutase (*sod*) gene sequences obtained from four species (representing three genera) of halophilic archaebacteria. These genes exhibit a pattern quite different from that outlined above. Before beginning the discussions, I will briefly describe some of the interesting features of halophilic archaebacteria and the origins of superoxide dismutase activity.

Halophilic archaebacteria: Based upon small subunit rRNA sequences and a number of other molecular sequences, living organisms divide into three distinct groups: eubacteria, eucaryotes, and the more recently recognized archaebacteria (also referred to as Bacteria, Eucarya, and Archaea, respectively).[20,21] One of the intriguing features of archaebacteria is their ability to propagate in extreme environments (i.e., anaerobic, extreme pH, hyperthermal and hypersaline) that are unfavourable for and usually lethal to most other organisms.[2,10] Halophilic archaebacteria both prefer and require two to five molar salt. Osmotic problems are avoided because these organisms maintain a balance between intra- and extracellular ionic strength. Equally importantly, the enzymes and other macromolecules of these organisms have evolved the capacity to maintain their structure and to function in high salt where electrostatic interactions are often masked and hydrophobic interactions are intensified.[10]

What are the available strategies for modifying enzymes so that they function in high salt environments? [3,10,22] It is well known that halophiles concentrate primarily K^+ cations inside the cell to counterbalance the extracellular cationic (predominantly Na^+) strength. Halophilic proteins avoid the salting out effects caused by high concentrations of KCl by substantially reducing surface hydrophobicity and increasing the content of acidic amino acid residues on the solvent exposed surface of the protein. These charges, which destabilize the protein in low salt, are effectively neutralized in high salt. The side chains of aspartic acid and glutamic acid in low salt solvent hydrate twice as much water as nonacidic residues.[17] But in high salt solvents, protein constituents cannot effectively compete directly with the highly hydrated ions for available water. Instead, the halophilic proteins are believed to have evolved a quaternary structure that coordinates hydrated salt ions to a higher local concentration than present in the surrounding solution.[3,10,22] The carboxyl groups located on the solvent exposed surface are believed to cooperatively nucleate and coordinate this network of water and salt ions that surround the protein. The core of the protein containing the more hydrophobic active site is conserved, protected, and structurally maintained. Physical measurements made on the malate dehydrogenase protein from *Haloarcula marismortui* have shown that the molecular particles in a solution of one to four molar KCl consist of 0.85 g of structured water and 0.35 g of salt per gram of protein.[22] Thus, the spatial distribution of acidic amino acid residues is probably a critical feature for maintaining surface hydration of halophilic proteins in high salt solvents.

Sod activity: The superoxide anion (O_2^-) is primarily a by-product of aerobic metabolism. This potentially toxic product is eliminated in a dismutation reaction catalyzed by SOD to produce molecular oxygen (O_2) and peroxide (O_2^{-2}),[5] which is rapidly eliminated by catalase and other peroxidases. The critically important SOD activity has appeared twice during evolution. Eucaryotic organisms possess a cytoplasmic enzyme that contains Cu^{+2} and Zn^{+2} as metal ion cofactors whereas most eubacteria and archaebacteria possess an enzyme that contains either Fe^{+2} or Mn^{+2} as metal ion cofactor. The amino acid sequences of the two types of SOD enzymes are unrelated (nonhomologous), although they catalyze the same dismutation reaction.

Halophilic sod *genes and the proteins they encode*: This work began with the purification of the SOD protein from *Halobacterium cutirubrum*.[11] The protein was shown to contain Mn^{+2} as the metal ion cofactor, to have optimal enzyme activity in buffers containing at least two molar salt, and to be homologous to the Fe^{+2} or Mn^{+2} SODs of eubacteria and unrelated to the CuZn enzymes of eucaryotes. The gene encoding this protein was subsequently cloned using degenerate oligonucleotides based upon the partial amino acid sequence of the purified protein.[12] Southern hybridization of the cloned *sod* gene sequence to restriction digests of genomic DNA revealed a second *sod*-like gene (*slg*) within the genome. This gene was cloned and characterized.[13] The comparison of the duplicated (paralogous) *sod* and *slg* genes revealed an inordinately high proportion of nonsynonymous nucleotide substitutions and suggested that the divergence of the two sequences was subject to intense selection.

To obtain additional information on the evolutionary history of *sod* genes in halophilic archaebacteria and to gain insight into the nature and extent of selection, a total of five additional *sod* genes were cloned and characterized from the closely related organism *Hb. sp. GRB* and the more distantly related organisms *Haloferax volcanii* and *Haloarcula marismortui*.[7,8] In three of the organisms, there are paralogous (or duplicated) *sod* genes whereas in *Ha. marismortui* there is only a single *sod* gene. The alignment of the seven genes is interrupted by gaps at two positions (Fig. 1). The first is the result of a deletion of codon three from the *sod2* gene of *Hf. volcanii*, and the second is the result of an insertion of nine nucleotides between codons four and five of the *Ha. marismortui* single copy *sod* gene. The other five genes encode proteins of 200 amino acids in length. As expected, all of the proteins contain a high proportion of acidic amino acid residues and have pI's between 4.16–4.20 (Table 1).

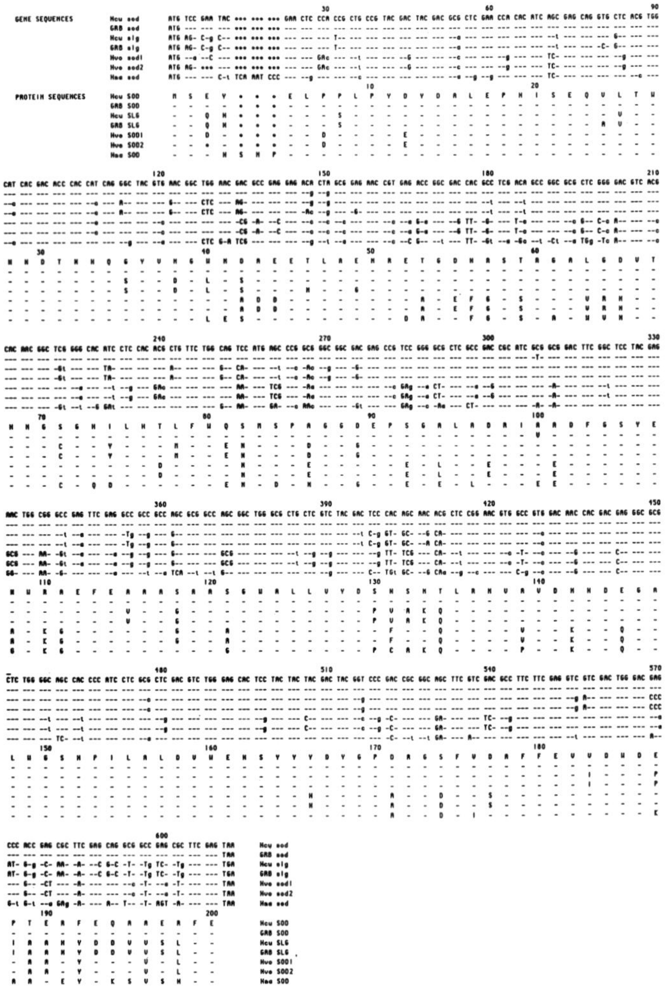

Figure 1. Alignment of *sod* gene nucleotide and amino acid sequences. Seven members of the *sod* gene family have been cloned and sequenced from four species: *Halobacterium cutirubrum* (*Hcu*), *Halobacterium spp GRB* (*GRB*), *Haloferax volcanii* (*Hvo*), and *Haloarcula marismortui* (*Hma*). Three of the species contains paralogous (duplicated) genes whereas the fourth species contains only a single *sod* gene. Symbols used are as follows: (-), nucleotide or amino acid identities when compared to the *Hcu sod* nucleotide or *Hcu* SOD amino acid sequence; (•), deletion of nucleotide (or amino acid) residues required to maintain alignment. In the nucleotide alignment, upper case substitutions are nonsynonymous and lower case substitutions are synonymous. The nucleotide numbering is above the gene sequences and includes all residue positions. The amino acid (or codon) numbering above the protein sequence ignores the three extra residues in the *Hma* SOD that occur between positions 4 and 5.

Table 1. Acidic amino acid residues in SOD-like proteins

			Number of acidic amino acid residues				
Protein[1]	Protein[2] Length	pI	Aspartate	Glutamate	Total	Mono-morphic	Poly-morphic
Hcu SOD	200	4.17	18	20	38	21	17
GRB SOD	200	4.16	18	20	38	21	17
Hcu SLG	200	4.17	16	20	36	21	15
GRB SLG	200	4.17	15	20	35	21	14
Hvo SOD1	200	4.17	21	18	39	21	18
Hvo SOD2	199	4.20	20	18	38	21	17
Hma SOD	203	4.16	18	23	41	21	20

[1]The sequence of the seven halophilic superoxide dismutase proteins were deduced from their gene sequences. Species abbreviations are: *Hcu*, *Hb. cutirubrum*; *GRB*, *Hb. sp. GRB*; *Hvo*, *Hf. volcanii* and *Hma*, *Ha. marismortui*.

[2]Protein length is in amino acid residues.

A phylogenetic tree derived from the nucleotide sequences of seven genes was constructed by neighbor joining (Fig. 2).[16] In this tree, the *Haloarcula* lineage appears to have diverged earliest with the *Haloferax* and *Halobacterium* lineages diverging shortly thereafter. The most recent speciation event was the splitting of *Hb. cutirubrum* from *Hb. sp. GRB*. A tree with identical topology was obtained using PAUP.[8]

The pairwise nucleotide and amino acid identities of the *sod* genes and the proteins they encode from the four halophilic organisms are summarized in Table 2. Three of the gene pairs in the matrix are greater than 99% identical. They are the orthologous *sod* and orthologous *slg* gene pairs from *Hb. cutirubrum* and *Hb. sp. GRB* and the paralogous *sod*1 and *sod*2 genes from *Hf. volcanii*. In contrast, the two other paralogous gene pairs, the *sod* and *slg* genes in *Hb. cutirubrum* and in *Hb. sp. GRB* are only about 87% identical. These observations suggest that the common ancestor of the *Haloferax* and *Halobacterium* lineages almost certainly had duplicated *sod* genes. In the *Haloferax* lineage, the coding regions of the paralogous genes have remained virtually identical, probably through a gene conversion-like mechanism. In the *Halobacterium* lineage, the paralogous genes have diverged substantially. The major portion of the divergence occurred prior to the species separation of *Hb. cutirubrum* and *Hb. sp. GRB*. Since speciation, the orthologous *sod* genes have accumulated three additional substitutions and the orthologous *slg* genes have accumulated five additional substitutions. The single *sod* gene in *Ha. marismortui* appears to be the most divergent and exhibits 76–78% nucleotide sequence identity to the other six genes. In all of the pairwise comparisons, there are an inordinately high proportion of nonsynonymous nucleotide substitutions. In most cases, the sequence identity between two proteins is less than that exhibited by the corresponding genes.

Protein structure and function: Three dimensional X-ray structures have been

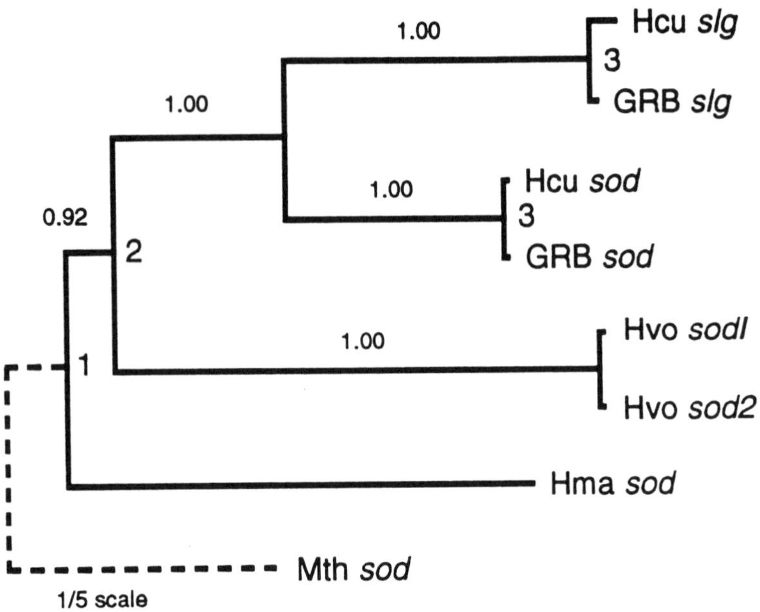

Figure 2. A phylogenetic tree of halophilic *sod* gene sequences. The tree was constructed using the neighbor joining method from the alignment of nucleotide sequences.[7,8,18] The *sod* gene sequence from *Methanococcus thermoautotrophicum* (*Mth*) was used as the out group to root the tree. Other abbreviations are as in Table I. Bootstrap values (100 repetitions) are indicated. The three speciation events are indicated: (1) divergence of the *Haloarcula* lineage, (2) divergence of the *Haloferax* and *Halobacterium* lineages and (3) divergence of *Hb. cutirubrum* from *Hb. sp. GRB*. The other two internal nodes represent the divergence of the paralogous *sod* and *slg* genes in the *Halobacterium* lineage and the paralogous *sod1* and *sod2* genes in *Hf. volcanii*.

solved for the Mn^{+2} SODs from *Bacillus stearothermophilus* and *Thermus thermophilus*.[18] By using these structures and alignments between the halophilic and eubacterial SOD protein sequences, it has been possible to identify by homology important structural and functional features and critical amino acid residues in the halophilic proteins.[8,15,19]

In the alignment of the seven gene sequences (Fig. 1), there are a total of 106 synonymous and 145 nonsynonymous nucleotide substitutions occurring at 123 codon positions (Table 3); there are 77 monomorphic codon positions that are devoid of substitutions. The distribution of these sites across the gene (or protein) is illustrated in Figure 3. Of the monomorphic codons, many specify critical amino acid residues that are responsible for features such as (i) binding of the Mn^{+2} metal ion cofactor, (ii) formation of the hydrophobic pocket around the active site, and (iii) contact between subunits in the homotetrameric structure.

Table 2. Percent Nucleotide and amino acid identities in halophilic sod genes and proteins

nuc[1] a.a.	Hcu sod	GRB sod	Hcu slg	GRB slg	Hvo sod1	Hvo sod2	Hma sod
Hcu sod	—	99.5 (0/1/2)	87.2 (25/21/31)	86.7 (25/24/31)	81.0 (25/28/61)	80.9 (26/29/59)	77.7 (36/35/63)
GRB sod	99.5	—	87.3 (25/22/29)	86.7 (25/25/30)	80.7 (25/29/62)	80.6 (26/30/60)	78.0 (36/34/62)
Hcu slg	82.5	82.0	—	99.2 (0/3/2)	76.3 (40/30/71)	77.1 (39/29/69)	77.2 (37/30/70)
GRB slg	81.0	80.5	98.5	—	76.0 (41/33/70)	76.7 (39/32/68)	76.3 (37/33/72)
Hvo sod1	80.0	79.5	72.0	70.5	—	99.5 (1/1/1)	76.3 (29/26/87)
Hvo sod2	80.4	79.9	72.4	70.9	100	—	76.2 (30/27/85)
Hma sod	74.5	75.0	76.0	74.5	78.0	78.4	—

[1]The abbreviations are the same as in Table 1. The percent nucleotide identity and the distribution of nucleotide substitutions between first, second and third codon positions is indicated above the diagonal. The percent amino acid identity is indicated below the diagonal.

Table 3. Nucleotide substitutions in the alignment of seven sod-like sequences

Nucleotide Substitions[1]	Number	Codon Position 1st	Codon Position 2nd	Codon Position 3rd	Codons Affected
synonymous	106	—	—	106	89
non-synonymous	145	64	58	23	77
Total	252	64	58	129	123

[1]The total number of nucleotide substitutions ocurring in the alignment of the seven halophilic *sod* genes are presented. The data have been tabulated from the alignments presented in Figure 1.

Twenty-one of the monomorphic codons specify either glutamic acid or aspartic acid in the protein. One of these (Asp158) is required for Mn^{+2} binding and another (Glu161) is involved in a critical subunit interaction. This important electrostatic interaction involving Glu161 is probably maintained in halophilic proteins because it is buried deep inside the homotetramer and well protected from ionic shielding effects. Some of the other 19 monomorphic acidic codons are undoubtedly involved in hydration at the surface of the protein.

Again, by homology with the known X-ray structure,[19] the probable location of α helix and β sheet elements within the halophilic SOD protein can be predicted (Fig. 3). In the eubacterial structures, the active site is surrounded by α helices 1, 2 and 6, β sheets 2 and 3, and the short region connecting β sheet 3 to α helix 6. In the halophilic proteins, the regions around three of the four metal

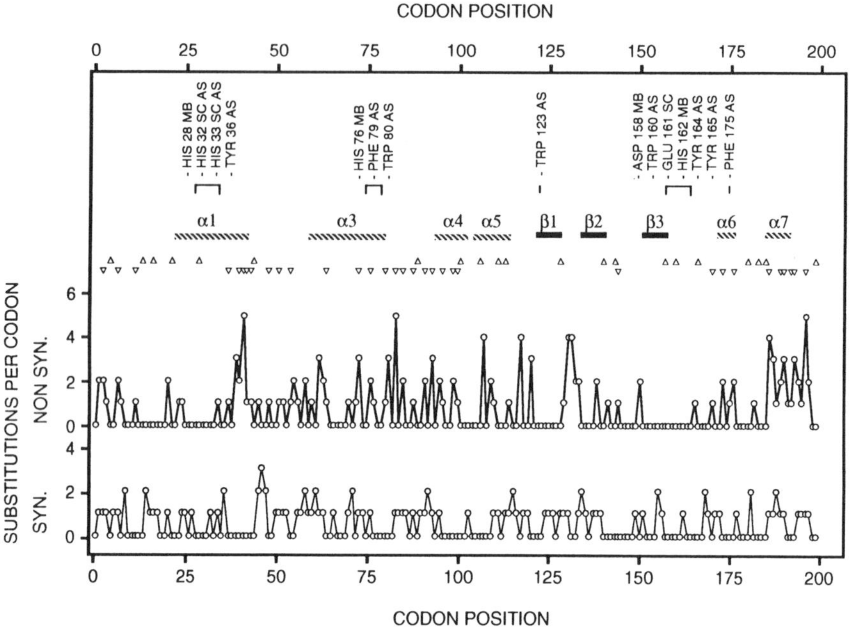

Figure 3. The distribution of synonymous and nonsynonymous substitutions in the seven halophilic *sod* genes. The two lower curves illustrate the number of synonymous and nonsynonymous substitutions occurring within each codon in the alignment of the seven halophilic *sod* genes. The downward pointing symbols (∇) represent positions in the proteins that are polymorphic and contain at least one acidic amino acid residue at that position. The upward pointing symbols (Δ) are monomorphic positions that contain either all aspartic acid or all glutamic acid residues at that positions. The putative locations of α helix and β sheet structural motifs deduced from X-ray structures of eubacterial Mn^{+2} SODs are indicated. The region corresponding to α helix 2 has been deleted from all of the halophilic proteins. The residues known to be of functional importance and conserved in the halophilic proteins are indicated according to their position. Abbreviations are: MB, residues involved in the binding of the Mn^{+2} metal ion cofactor; SC, residues involved in subunit contact within the homotetrameric enzyme structure; AS, residues that are near and participate in the formation of the hydrophobic active site of the enzyme.

binding residues (His28, Asp158, and His162) are highly conserved; these regions contain many of the residues believed to be critical for enzyme function.[8]

The 145 nonsynonymous nucleotide substitutions are confined to just 77 codons. Of these polymorphic positions, 33 contain one or more codons that specify acidic amino acid residues at the alignment position in the seven proteins. This high proportion of nonsynonymous substitutions indicates that these proteins are subject to continuous evolutionary tinkering resulting from fluctuating selection pressures. The high proportion of replacements involving acidic residues implies that a substantial proportion of this tinkering involves the distribution of acidic amino acid residues and by inference the adaptation of the proteins for function in high salt.

Serine codon utilization: Also of interest is the utilization of serine codons in the halophilic *sod* genes.[1,7] Serine can be specified by either the TCN or AGY codon families. In halophiles, there seems to be no restriction as to which family can be used although in either case, because of the base composition of genomic DNA, G or C is found preferentially at the third position.[18]

Briefly, there are 16 alignment positions where serine occurs in at least three of the seven proteins. At nine of these positions, listed in Table 4, both families of serine codon are employed. At the three positions where only serines are present (positions 2, 21 and 151) at least two nonsynonymous substitutions are required to connect the codons. This implies that in some ancestral sequences, an amino acid other than serine must have been present at this position. In only one instance (position 59), are the two serine codon families linked by an intermediate and related threonine codon. Again, two nonsynonymous substitutions are required to connect the different serine codons. In all other cases (positions 42, 84, 188, 132 and 197), at least three nonsynonymous substitutions are required to connect all codons. These results can be explained in two ways: either there has been convergent evolution where the two types of serine codons

Table 4. Positions that use both the TCN and AGY families of serine codon

		Codon position[1]								
		2	**21**	**42**	**59**	**84**	**118**	**132**	**151**	**197**
	aa	S	S	D	T	S	S	S	S	E
Hcu sod	codon	TCC	AGC	GAC	ACA	AGC	AGC	AGC	AGC	GAG
	aa	S	S	D	T	S	S	S	S	E
GRB sod	codon	TCC	AGC	GAC	ACA	AGC	AGC	AGC	AGC	GAG
	aa	S	S	S	T	S	G	A	S	S
Hcu slg	codon	AGC	AGT	AGC	ACT	ACT	GGC	GCC	AGC	TCG
	aa	S	S	S	T	S	G	A	S	S
GRB slg	codon	AGC	AGT	AGC	ACT	AGT	GGC	GCC	AGC	TCG
	aa	S	S	A	S	S	G	S	S	E
Hvo sod1	codon	TCA	TCC	GCG	TCC	TCG	GGC	TCG	AGC	GAA
	aa	S	S	A	S	S	G	S	S	E
Hvo sod2	codon	AGC	TCC	GCG	TCC	TCG	GGC	TCG	AGC	GAA
	aa	S	S	S	S	D	S	A	S	S
Hma sod	codon	TCC	TCC	TCG	AGC	GAC	TCA	GCC	TCC	AGT

[1]Codon positions in the seven halophilic *sod* genes that use both the TCN and AGY serine codons are listed. The specified amino acid is indicated above each codon: S, serine; D, aspartic acid; A, alanine; T, threonine; G, glycine; E, glutamic acid. Codons specifying amino acids other than serine are boxed.

were derived independently at a given position in two different ancestral sequences that had already diverged, or alternatively the amino acid position specified by these codons may be very fluid and subject to fluctuating selection. Of the six serine positions in Table 4 that exhibit amino acid polymorphism, three of the replacements involve at least one aspartic or glutamic acid residue. These residues are likely candidates for selection because of their role in surface hydration of halophilic protein.[22]

Flanking sequences: The seven halophilic *sod* genes are all transcribed predominantly as monocistronic mRNAs from short (TTAA) promoter elements located immediately upstream of the ATG translation initiation codons.[7,8,12,13] Transcription termination occurs at a short poly T track that is positioned between 20 and 150 nucleotides beyond the translation termination codon in the different genes. Two distinct types of regulation are seen for the paralogous gene pairs. The *sod* genes of *Hb. cutirubrum* and *Hb. sp. GRB* and the *sod*1 gene of *Hf. volcanii* are inducible (as measured by enzyme activity and mRNA accumulation) by the drug paraquat. Paraquat elevates the level of the superoxide anion (O_2^-) by interrupting the electron transport chain. The *slg* genes of *Hb. cutirubrum* and *Hb. sp. GRB*, the *sod*2 gene of *Hf. volcanii* and the single copy *sod* gene of *Ha. marismortui* are transcribed constitutively and are not induced by paraquat.

In the entire set, the only two gene pairs that exhibit substantial sequence identity in the 5′ and 3′ flanking regions are the recently diverged orthologous *sod* and orthologous *slg* genes of *Hb. cutirubrum* and *Hb. sp. GRB*.[7] For the two gene pairs, a total 533 nucleotides of common 5′ and 3′ flanking sequence have been determined and only three nucleotide substitutions were found (Table 5). In the 1200 nucleotides of gene sequence, three substitutions were found in the *sod* genes and five substitutions were found in the *slg* genes. Of these eight substitutions, three are synonymous, four are nonsynonymous, and one is ambiguous. One of the nonsynonymous substitutions replaces an otherwise conserved glutamic acid with glycine at position 56 in the *slg* protein of *Hb. sp. GRB*. These results indicate that divergence within and between the *sod* and *slg* genes has continued in the Halobacterium lineage after the separation of *sp. cutirubrum* and *sp. GRB*. In fact, the substitution rate in the coding regions of the *sod* and *slg* gene is nearly twice that observed in the flanking regions (Table 5).

Surprisingly, there is no similarity between the sequences flanking the paralo-

Table 5. Nucleotide substitutions in coding and flanking regions of the orthologous *sod* and *slg* genes of *Hb. cutirubrum* and *Hb. sp. GRB*.

	coding		flanking			
	sod	slg	5′ sod	5′ slg	3′ sod	3′ slg
substitution	3/600	5/600	1/174	0/60	0/239	1/60
total	8/1200 = 0.7%		2/533 = 0.4%			

gous *sod* and *slg* genes in either *Halobacterium* species or between the sequences flanking the paralogous *sod*1 and *sod*2 genes of *Hf. volcanii.*[7] In the first two cases, the coding regions are about 87% identical and in the third case the coding regions are 99.5% identical. This implies that the duplication event is either very ancient and that any similarity in the flanking regions that once existed has been totally obliterated or that the duplication covered only the coding region of the gene.

It was anticipated that *sod* genes that are inducible by paraquat might exhibit some sequence similarity in their 5′ flanking regions that reflected their common response to paraquat. Indeed, the 5′ flanking region of the *sod* genes from *Hb. cutirubrum* and *Hb. sp. GRB* are identical but this is because of their recent speciation. When they were compared to the 5′ flanking region of the paraquat inducible *sod*1 genes of *Hf. volcanii*, no similarity was observed.[7] Given the absence of sequence similarity, it is difficult to explain regulation by paraquat.

The *slg* genes of *Hb. cutirubrum* and *sp. GRB*, the *sod*2 gene of *Hf. volcanii* and the *sod* gene of *Ha. marismortui* are not inducible by paraquat. Surprisingly, the 5′ flanking regions of the latter two genes exhibit about 50% nucleotide sequence similarity to each other and to the *slg* genes of *Halobacterium* over a region of 60 nucleotides encompassing the TTAA promoter element.[7] The significance of this limited sequence similarity is not known.

The divergence of another halophilic gene: Halophilic bacteria contain gas vacuoles that are used to regulate buoyancy in hypersaline environments.[4] The *vac* gene encodes the major structural protein of the gas vacuole. The *vac* gene is present in one copy per genome in *Hf. mediterraneii* and two (paralogous) copies per genome in *Hb. halobium* (the same species as *Hb. cutirubrum*). These genes have been cloned and sequenced.[4] Comparisons indicate that in the *vac* genes, synonymous substitutions are about tenfold more frequent than nonsynonymous substitutions. These data indicate that at least some sequences within halophilic archaebacteria diverge according to the more standard models.

Conclusion

The divergence of the seven members of halophilic *sod* gene family appears to be driven by intense but fluctuating selection. It is unlikely that selection has been used to generate a protein with a novel enzymatic activity because virtually all of the critical residues responsible for the ancestral superoxide dismutase activity are conserved in each of the seven halophilic proteins.[8,15,19] It is more likely that selection is adjusting and refining the structure of the protein to make it function more efficiently in high salt environments. A number of observations support this view. First, the number of nonsynonymous substitutions occurring in the seven sequences is nearly twice as great as the number of synonymous substitutions. Examination of a large number of other sequences including the *vac* genes in halophilic archaebacteria indicates that synonymous substitutions are usually five to twenty-fold more prevalent than nonsynonymous substitutions.[4,14] Secondly, many of the nonsynonymous substitutions result in replacements involving glutamic acid or aspartic acid. Acidic residues are known to play an important role in adapting proteins for function in high salt.[2,3,10,22] It is believed that they distribute along the exposed surface of the protein, and cooperatively nucleate and coordinate a network of water and salt ions that surround the protein. Thirdly, the majority of alignment positions at which serine occurs in at least three of the sequences, both the TCN and AGY codon families are used. The simplest explanation for this is that these sites are being subjected to evolutionary tinkering on a continuous basis in order to refine or optimize activity in high salt environments. Fourthly, comparison of the recently diverged orthologous *sod* and orthologous *slg* genes from *Hb. cutirubrum* and *Hb. sp. GRB* indicates that the coding sequences are continuing to accumulate nonsynonymous substitutions in excess of synonymous substitutions.[7] One of the substitutions results in the replacement of a highly conserved glutamic acid residue with glycine. Finally, the substitution rate within the genes is nearly twice as great as the substitution rate in the 5′ and 3′ flanking regions.

Acknowledgments

This work has been supported by a grant from the Medical Research Council of Canada. PPD is a fellow of the Canadian Institute for Advanced Research, Program in Evolutionary Biology. I wish to thank Simon Potter, Phalgun Joshi, and Bruce May for their interest in this work. I also thank Moshe Mevarech for sharing his insights relating to structural features of halophilic proteins.

References

1. S. Brenner. (1988). *Nature*, 334, 428–430.
2. P. Dennis. (1986). *J. Bact.* 168, 471–478.
3. H. Eisenberg, M. Mevarech, and G. Zaccai. (1992). *Advan. Prot. Chem.* 43, 1–62.
4. C. Englert, M. Horne, and F. Pfeifer. (1990). *Mol. Gen. Genet.* 222, 225–232.
5. I. Fredovich, 1986. *Adv. Enzymol.* 58, 68–97.
6. R. R. Hudson, M. Kreitman, and M. Aquade. (1987). *Genetics* 116, 153–159.
7. P. Joshi, and P. P. Dennis. (1992). *J. Bacteriol.* 175, 1561–1571.
8. P. Joshi, and P. P. Dennis. (1992). *J. Bacteriol.* 175, 1572–1579.
9. M. Kimura. (1986). *Nature* 217, 624–626.
10. J. Lanyi. (1974). *Bacteriol. Rev.* 38, 272–290.
11. B. P. May, and P. P. Dennis. (1990). *J. Bacteriol.* 172, 3725–3729.
12. B. P. May, and P. P. Dennis. (1987). *J. Bacteriol.* 169, 1417–1422.

13. B. P. May, and P. P. Dennis. (1989). *J. Biol. Chem.* 264, 12253–12258.

14. H. Ochman, and A. C. Wilson. (1987). *J. Mol. Evol.* 26, 74–86.

15. M. Parker, and C. C. Blake. (1988). *Febs. Letters* 119, 377–382.

16. N. Saiton, and M. Nei. (1987). *Mol. Biol. Evol.* 4, 406–425.

17. W. Seanger. (1987). Ann. Rev. Biophys. Chem. 16, 93–114.

18. L. Shimmin, and P. P. Dennis. (1989). *EMBO J.* 8, 1225–1235.

19. W. C. Stallings, K. A. Partridge, R. K. Stong, and B. Ludwig. (1985). *J. Biol. Chem.* 260, 16424–16432.

20. C. R. Woese. (1987). *Microbiol. Rev.* 51, 221–271.

21. C. R. Woese, O. Kandler, and Wheelis. (1990). *Proc. Natl. Acad. Sci. USA* 87, 4576–4579.

22. G. Zaccai, F. Candrin, Y. Haik, N. Borokov, and H. Eisenberg. (1989). *J. Mol. Biol.* 208, 491–500.

15

Mitochondrial Haplotype Frequencies in Oysters: Neutral Alternatives to Selection Models

Andrew T. Beckenbach

Introduction

The animal mitochondrial genome is a small circular molecule of about 15–42kb (kilobase pairs). Although present in multiple copies in each cell, it is essentially haploid and its inheritance is predominantly maternal. The molecule undergoes very rapid evolution at the DNA sequence level, due in part to the apparent lack of repair during replication. Because of its simple mode of inheritance, its ease of extraction and analysis and the extensive variation that can be observed in most animal populations, mitochondrial DNA (mtDNA) has been the focus of many population surveys since the late 1970s.[1,2]

Initial surveys, primarily involving mammals, showed that most of the intraspecific variation occurred between populations.[3] Most individuals within populations carry a single common haplotype, with perhaps one or two rare variants. Further surveys, including invertebrates and non-mammalian vertebrate species, showed that many species have considerably more population level variation than was initially found, usually including one or a few relatively common haplotypes, some of intermediate frequency and a number of rare variants.[2] Each of these frequency distributions can be explained on the basis of neutral expectation using Ewens' sampling theory.[4,5] In fact, they represent the same distribution, differing only in the parameter, θ. For small values of θ, a single haplotype will predominate, with only a few variants. For large θ, many more haplotypes are expected, with none occurring at high frequency.

The parameter, θ, is actually a composite of two parameters, μ, the rate at which new, unique neutral mutations arise, and N_e, the effective populations size.[4] These two parameters affect the haplotype distributions in slightly different ways. Mutation introduces new variation into the population at initial frequencies of 1/N, where N is the observed population size, while N_e determines the rate which drift alters the frequencies of existing variation, including the elimination of rare haplotypes. In populations having small values of N_e, the majority of individuals carry a single common haplotype. These two parameters are normally confounded, and their separate effects cannot be distinguished from haplotype frequency distributions.

Some Oyster Reproductive Biology

Oysters are protandrous (i.e., sequential) hermaphrodites that live in extensive beds in shallow marine water.[6] The adults live for a number of years, and in B.C. are able to breed successfully in only the most favorable years. Breeding adult males and females release gametes into the water column where fertilization takes place. The larvae are planktonic and many are lost due to predation, disturbance of the water column or currents washing them from suitable benthic habitats. Only a very small proportion of larvae find suitable substrates. The oysters reach reproductive maturity first as males and release large numbers of sperm at the time of spawning. Each year, a proportion of reproductive males transform into females.

Several generalizations can be drawn from oyster life history. (1) The actual population sizes of oyster beds (N) is quite large. (2) To compensate for the enormous wastage of gametes in the planktonic stages, each breeding female produces an enormous number of eggs, sufficient to replace the entire adult population. Individual females may produce as many as 90 million eggs at a single spawning, and may spawn more than once a year.[6] Thus the variance in family size may be extremely large, and the effective population size (N_e) may therefore be quite small. (3) The germline tissue of a breeding female, which reproduces first as a male, producing large numbers of sperm cells, and later as a female generating large numbers of eggs, must undergo a very large number of cell divisions per animal generation. In contrast, most animal species minimize the number of germline cell divisions per animal generation, presumably to minimize the accumulation of mutations. In oysters, the extensive production of gametes demanded by this life history characteristic may lead to a rather high frequency of mutations compared to other animal species.

Haplotype Distributions in Oysters

Restriction analysis of mtDNA from the Pacific oyster (*Crassostrea gigas*) produced a haplotype frequency distribution quite different from those usually observed in such surveys.[7] Forty-four haplotypes were observed, yet a very high proportion of individuals fell into only two haplotype classes. The rank order frequency distributions observed and those expected under Ewens' sampling theory are given in Figure 1. Haplotype frequency distributions of the American

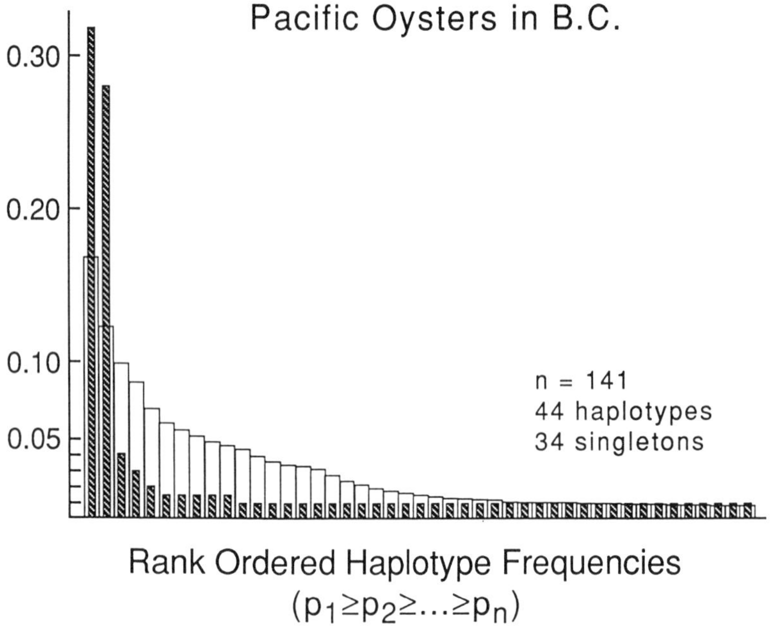

Figure 1. Haplotype frequencies of *Crassostrea gigas* in B.C. Central, dark bars are observed values; open bars are frequencies expected according to Ewens sampling theory. Data from Boom, et al., in press.

oyster (*C. virginica*) for two genetically differentiated geographical regions are given in Figure 2.[8] In each of these distributions there appears to be an excess of the common haplotype(s), a deficiency of haplotypes at intermediate frequencies and too many singletons (haplotypes observed only once). The Ewens-Watterson test[9] provides a conservative method for determining whether these deviations are significant. The test statistic is $F = \Sigma p_i^2$, where p_i is the frequency of the ith haplotype in each population. This statistic is the inverse of the effective number of alleles (n_e), so for large numbers of haplotypes, F is expected to be small. For the three distributions in Figures 1 and 2, F = 0.188, 0.205 and 0.432 for the B.C., Gulf and Atlantic populations. There are no tabulated values for significance for large numbers of haplotypes and large sample sizes, so it is necessary to simulate distributions with the values observed in each population under neutral expectations.[10,11] In 2000 simulations for each of the three populations, none of the values of F from the simulated populations exceeded the test values. Thus the observed distributions deviate significantly from neutral expectation ($p < 0.0005$).

Why Not Selection?

The null hypothesis in the Ewens-Watterson test is usually regarded as neutrality, and rejection of that hypothesis is taken as evidence of selection. There are, however, other assumptions that are part of Ewens' sampling theory and are therefore part of the null hypothesis. In particular, it is assumed that μ is small and N_e is large, so that θ is of intermediate magnitude. Because of the life history patterns noted above, we must consider the possibility that none of these assumptions hold for oyster populations. Perhaps more importantly, it is likely that the adult population sampled (N) is very much larger than the effective size (N_e), because of the enormous reproductive potential of individual females, and the low probability of success for each spawning female. Reproduction may be described as a lottery, with a few big winners and many losers. Samples for genetic studies cannot distinguish reproductive success and reflect the variation present in the adult population (N, not N_e).

As noted above, a majority of individuals fall into only one (Figure 2) or two (Figure 1) haplotypes. Predominance of one or two haplotypes could be explained either by strong purifying selection or small N_e. The difficulty with either of these explanations is that more than 30% of individuals in each population carry rare haplotypes. Very large populations under purifying selection can harbor

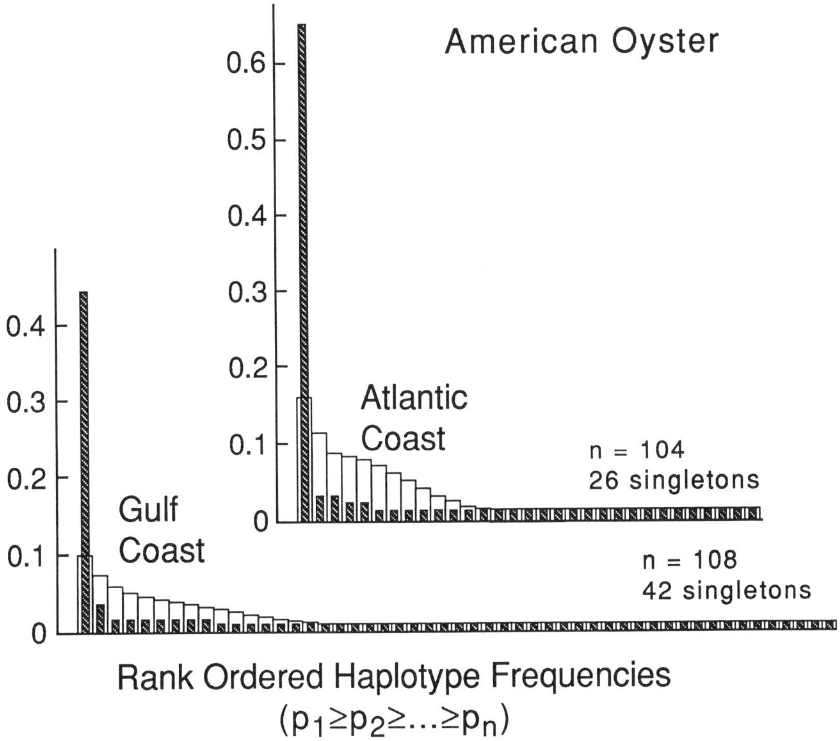

Figure 2. Haplotype frequencies of *Crassostrea virginica*. The populations from the Gulf of Mexico and those from the Atlantic Coast (inset) represent genetically distinct groups. Data from Reeb and Avise, 1990.

large numbers of deleterious recessive alleles for nuclear genes. It should be remembered, however, that mtDNA is effectively haploid and all variants are exposed to selection, if selection is present. It is very difficult to reconcile the strong purifying selection required to concentrate most individuals into one or two haplotypes with the presence of 34–56% of the members of the populations carrying rare variants.

A second reason to look beyond selection for an explanation of oyster haplotype frequencies comes from comparisons with surveys of mtDNA variation in other animal species, and an examination of the nature of variation exposed by restriction analysis. Restriction sites are scattered throughout the genome, and endonucleases sample both coding and non-coding regions. When variation is observed in intraspecific analyses, most of the variants in protein coding regions are silent transitions.[12,13] The fact that observed intraspecific variation is essentially confined to degenerate third codon positions, or to the non-coding control region, suggests that changes altering the function of the genes are subject to strong purifying selection. Yet we would not expect the silent variation that remains to be subject to strong purifying selection as well. The empirical observation from other animal groups is that where variation is observed (usually when N_e is apparently large) analysis of the variation at the sequence level shows that the majority of variants are silent and should approximate neutrality.

A Cladistic View of Oyster Haplotypes

One feature of oyster haplotypes is not apparent in the frequency distributions: the relationships among the variants. The two common haplotypes in the B.C. populations differ by a single restriction site.[7] The common haplotypes of the Gulf and Atlantic coast American oysters differ by restriction patterns at eight of 13 polymorphic enzymes. The original papers give UPGMA trees for the observed variants, but a weakness of this technique is that it produces a bifurcating tree. No bifurcating tree can adequately represent the relationships among these haplotypes.

Figures 3 and 4 give cladograms inferred from Table 1 of the two data papers.[7,8] The most striking feature of these cladograms is the relative lack of "networking" in the diagrams. The common haplotypes occupy central positions, with most of the rare haplotypes appearing as single steps from one of the common haplotypes. It should be noted that no frequency information was used in generating these cladograms: the common haplotypes have connections to many rare variants, while most of the rare variants have only a single connection. The common haplotypes are forced into central positions.

Over the past decade there has been an increasing interest in using cladistic approaches to the analysis of population variation. Several results of coalescent theory provide insight into these observations.[14] (1) The probability that a haplo-

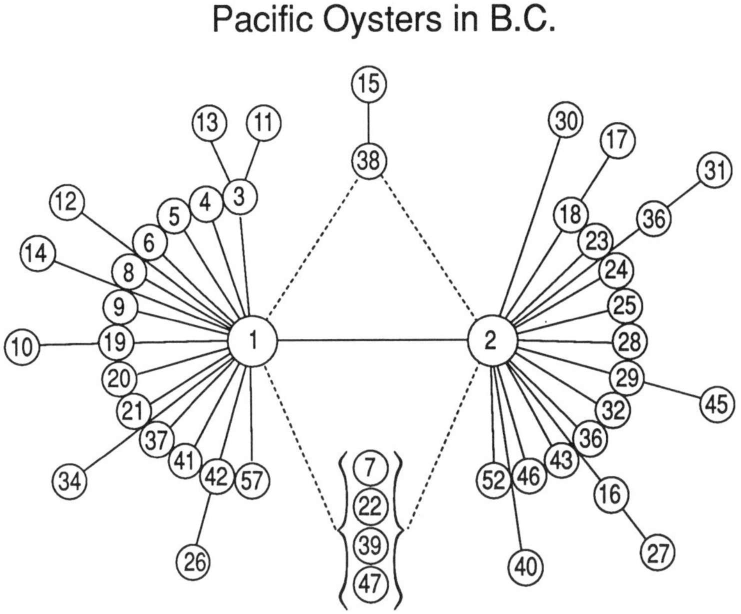

Figure 3. Cladogram of B.C. oyster haplotypes. Haplotypes 1 and 2 are the two abundant haplotypes, and differ by a single restriction site. The closest rows of circles differ from the associated common haplotype by the restriction pattern generated by a single restriction enzyme. The haplotypes that are ambiguous in their relationship to haplotypes 1 and 2 are indicated by dashed lines.

type is the oldest in the population equals its frequency in the population.[15] (2) The most recent haplotypes occur preferentially at the tips of the cladogram. (3) The number of mutational connections to a particular haplotype is related to frequency. (4) The probability a singleton is derived from allele i is p_i. All of these results, when applied to the frequency distributions and cladograms (Figures 1–4) imply that the common haplotypes are old and the rare ones are recently derived from them. There do not appear to be any haplotypes of intermediate age. This point is important, since it suggests that very few haplotypes survive in the long term; virtually all of the others are apparently transient and must be continuously replaced.

A few assumptions of coalescent theory should be kept in mind. (1) The analysis assumes an infinite alleles model. The oyster analyses sample only about 60 to 100 restriction sites. With as many as 82 observed variants (in the Gulf and Atlantic oysters combined) the possibility of multiple origins of haplotypes cannot be discounted. (2) Sampling theory requires that the sample be very much smaller than N_e. This assumption is required to eliminate the possibility of

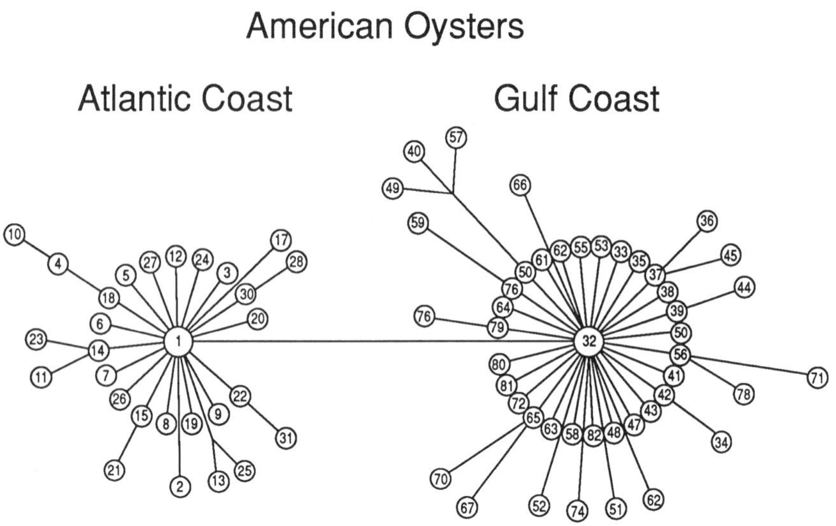

Figure 4. Cladogram of American oyster haplotypes. Haplotypes 1 and 32 are the abundant haplotypes in the Atlantic and Gulf Coast populations. They differ by restriction patterns generated by eight of thirteen polymorphic restriction enzymes. Data from Reeb and Avise, 1990. A few haplotypes could not be distinguished in their Table 1 and are omitted.

multiple coalescent events and multiple mutational events at each generation. Biological considerations suggest that oyster population effective sizes may be quite small due to large variance in reproductive success. These assumptions will be reexamined below.

Simulations

I undertook a series of computer simulations to recreate the major features of oyster haplotype distributions. The following population parameters could be specified for each run: actual population size (N), maximum reproductive potential for each female and rate of mutation to new neutral haplotypes (μ). By varying the reproductive potential for reproducing females, it was possible to vary the effective population size from about 4 females to 1.2N females. If a reproductive potential of 1 was defined, females were sampled with replacement and produced a single offspring. The result is a Poisson distribution of family size and $N_e \simeq N$. If a reproductive potential of $1 < x \leq N$ was chosen, females were sampled without replacement and the number of offspring was determined by drawing a random number on the interval [1,x]. When $x = 2$, the effective size was greater than N. When $x > 4$, $N_e < N$ and when $x = N$, N_e was observed to be about 4, regardless of N.

At each generation, females were drawn at random until N offspring were produced. Each offspring was subjected to mutation with probability μ. For each new haplotype, the generation of its origin was recorded so that the relative ages of common and rare haplotypes could be compared. At specified time intervals (usually every 100 generations), the number of haplotypes and their frequencies were recorded, and N_e and n_e (effective number of alleles, $1/\Sigma p_i^2$) were calculated and recorded. n_e is a measure of evenness of haplotype frequencies and equals the actual number of haplotypes (n) if all haplotypes are equally frequent in the population. If haplotype frequencies are very uneven as in Figures 1 and 2, $n_e << n$, and may approach 1. For oyster populations, $n_e = 1/F$ are 5.33, 4.88 and 2.31 while n = 44, 51 and 31 respectively for B.C., Gulf and Atlantic coast populations.

In the simulations, I searched for conditions under which large numbers of haplotypes were present in populations with relatively small values of n_e. I focus on comparatively small effective population sizes, which are required to concentrate most individuals into one or two haplotypes. Figure 5 shows the effects of N_e and μ on n and n_e for moderately large populations (N = 5000). For $N_e < 200$, n_e is small regardless of mutation rate. For populations with small N_e, exceedingly high mutation rates are required to generate large numbers of haplotypes. Figure 6 shows the effect of N on n and n_e for $\mu = 1/100$. For $N_e < 300$, N must be more than 10 times greater than N_e to include as many rare haplotypes as were observed in the oyster populations. Examination of the generation of origin of the rare haplotypes showed that nearly all arose in the preceding generation. That is, the majority of singletons survive only one generation.

For values of N_e larger than 500, the haplotype distributions approach those expected under Ewens' sampling theory, regardless of μ. Thus the simulations suggest that the two parts of the distributions in Figures 1 and 2 represent the separate effects of very small N_e, concentrating individuals into two or one haplotype, and very high μ, replenishing the rare mutants lost in each generation.

Assumptions of the Coalescent, Revisited

Two assumptions required in the application of coalescent theory appear not to apply to oyster mtDNA haplotypes: infinite alleles and relatively large N_e. About one-third of the members of each population appear to carry relatively new mutations distinguishable as restriction site differences. Only 60 to 100 restriction sites were analyzed in the two surveys.[7,8] If mutations are as common as hypothesized here, restriction patterns occurring in two or three individuals may actually reflect separate events. As well, it should be noted that 29 of the rare haplotypes in B.C. differ from the common forms by pattern at only one restriction enzyme (Figure 3), while 14 differ at two and three differ by three patterns. Similar

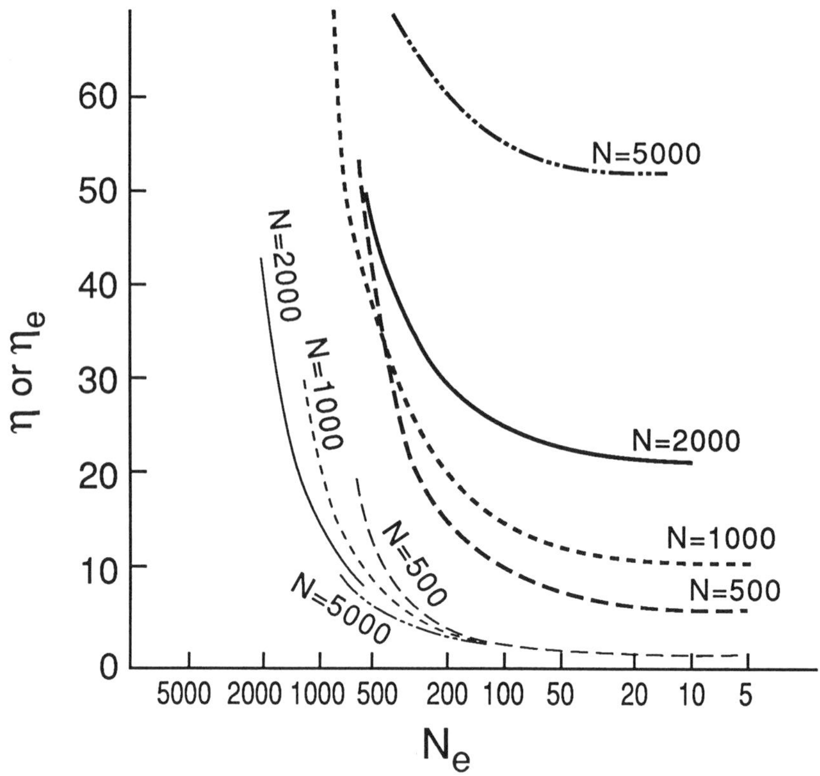

Figure 5. Simulation results: effect of mutation rate on the number of haplotypes (n; heavy lines in upper right) and the effective number of haplotypes ($n_e = 1/\Sigma p_i^2$; thin lines, lower left). Actual population size is N = 5000.

results hold for the American oyster populations, as well (Figure 4). If most of the singletons arose in the preceding generation, as appears necessary from the simulation results, then we may account for some of the two and three step events as multiple events occurring in individual germline lineages.

Conclusions

The frequency patterns from Pacific and American oyster populations (Figures 1 and 2) differ significantly from neutral expectations in two essential ways: the common haplotypes are far too common for the number of haplotypes in each population, and there are far too many singleton haplotypes. In the Pacific oysters, the two common haplotypes differ by only a single restriction site, and all the rare haplotypes are closely related to the common haplotypes (Figure 3). A similar result occurs in the American oyster. Although the Gulf and Atlantic

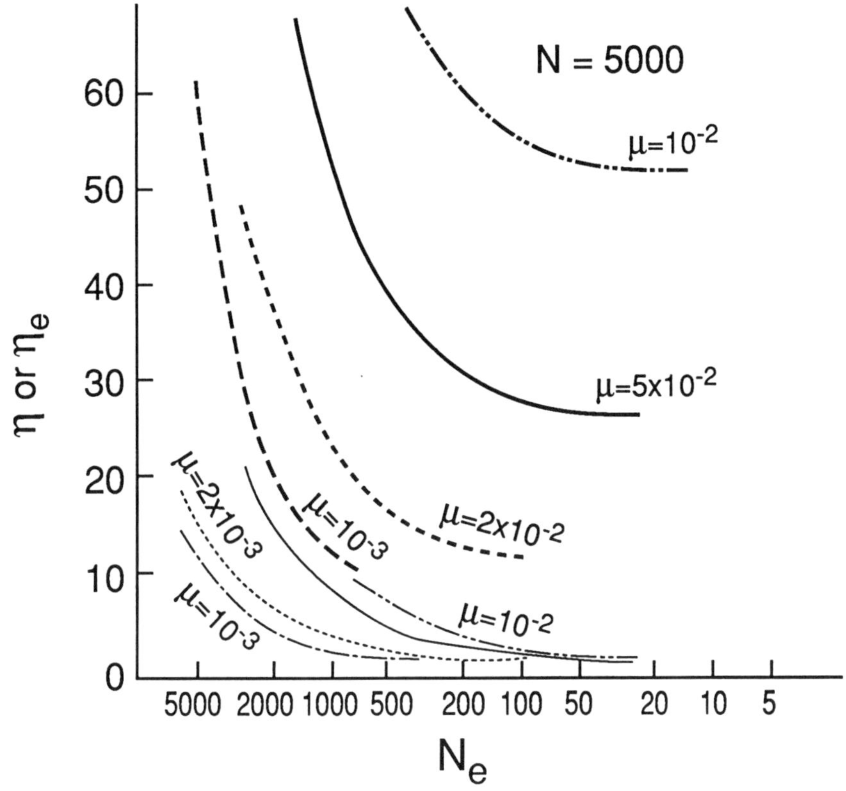

Figure 6. More simulations: The disparity between the effective number of haplotypes (n_e; thin lines, lower left) and the actual number (n; thick lines, upper right), increases with adult population size (N).

populations differ substantially, within each population the rare alleles appear very closely related to the associated common haplotype. I propose that these two characteristics of the distributions are manifestations of the separate effects of very small N_e, concentrating most members of the populations into the common haplotype(s) and very high μ, replenishing the rare haplotypes lost each generation.

One additional possibility must be examined: the assumption that the populations are at equilibrium. If the populations have undergone very recent expansion, it is possible that results similar to those observed might arise.[5] The B.C. populations are, in fact, recent transplants from Japanese populations. The American oysters, however, show similar frequency distributions. There is no reason to assume that these results occurred as a result of parallel non-equilibrium events in three different populations, B.C., Gulf and Atlantic coasts. More likely it is a product of life history characteristics of the oysters. The most important characteristics are the reproductive strategies of all oyster species: they are protandrous hermaphrodites, reproducing first as males then later as females, and are capable of producing hundreds of millions of eggs. Thus the potential variance in family size of individual females is enormous, suggesting very small effective population sizes. On the other hand, the enormous wastage of gametes, first as males then as females, suggests that many germline cell divisions must be required each generation. Thus the opportunity for accumulation of new mutations must be much higher than for other animal species. These two characteristics, very small N_e and very high μ, may account for the unusual haplotype distributions.

Acknowledgements

I am indebted to Monte Slatkin who suggested the cladistic approach to examining the haplotype distributions, and Chip Aquadro who suggested that recent population expansion can result in non-neutral distributions. This work was supported by a Research Grant from NSERC.

References

1. C. Moritz, T. E. Dowling, and W. M. Brown. (1987). *Annu. Rev. Ecol. Syst.* 18,269–292.
2. J. C. Avise, J. Arnold, R. M. Ball, E. Bermingham, T. Lamb, J. E. Neigel, C. A. Reeb, and N. C. Saunders. (1987). *Annu. Rev. Ecol. Syst.* 18,489–522.
3. J. C. Avise, and R. A. Lansman. (1982). in *Evolution of Genes and Proteins* (M. Nei, and R. K. Koehn eds.) pp. 147–164, Sinauer Associates.
4. W. Ewens. (1972). *Theor. Pop. Biol.* 3,87–112.
5. T. S. Whittam, A. G. Clark, M. Stoneking and R. L. Cann. (1986). *Proc. Natl. Acad. Sci. UXA* 83,9611–9615.
6. M. F. Strathmann. (1987). *Reproduction and Development of Marine Invertebrates of the Northern Pacific Coast*, U. Washington Press.
7. J. D. G. Boom, E. G. Boulding, and A. T. Beckenbach. (1994). *Can. J. Fish. Aquat. Sci.* (in press).
8. C. A. Reeb, and J. C. Avise. (1990). *Genetics* 124,397–406.
9. G. A. Watterson. (1978). *Genetics* 88,405–417.
10. F. M. Stewart. (1977). Appendix to P. A. Fuerst, R. Chakraborty, and M. Nei. *Genetics* 86,455–483.
11. B. F. J. Manly. (1985). *The Statistics of Natural Selection on Animal Populations*, Chapman and Hall.
12. A. T. Beckenbach, W. K. Thomas, and H. Sohrabi. (1990). *Genome* 33,13–15.
13. A. T. Beckenbach, Y. W. Wei, and H. Liu. (1993). *Molec. Biol. Evol.* 10,619–634.
14. K. A. Crandall, and A. R. Templeton. (1993). *Genetics* 134,959–969.
15. P. Donnelly, and S. Tavaré. (1986). *Adv. Appl. Probab.* 18,1–19.

16

Gene Duplication, Gene Conversion and Codon Bias

Donal A. Hickey, Shaojiu Wang and Charalambos Magoulas

Introduction

Within many genes, synonymous codons are found in very unequal frequencies. This observation of codon bias is often interpreted as a reflection of natural selection acting to mold the codon "choice" to match the frequency of corresponding tRNAs.[1–6] While the observation of codon bias is undeniable, the selective interpretation has been questioned. For instance, the fact that many different codons are biased toward the same nucleotide has led to the suggestion that codon bias may simply be a reflection of mutational bias.[7,8] On the other hand, arguments against mutational bias are based on the observation that the level of codon bias can vary dramatically between different genes in a single genome,[5] and between coding sequences and adjacent non-coding sequences.[9] In this paper we consider the possibility that at least some cases of non-random codon usage are due to a bias in DNA repair. Furthermore, we point out that DNA repair enzymes do not affect all genes in a uniform manner. For instance, the effects of DNA repair are especially pronounced in duplicated genes that are undergoing concerted evolution, and such genes do show an extreme bias in the distribution of synonymous codons. In general, competing theories about the causes of codon bias can be tested based on the predictions they make about the patterns of long-term evolutionary trends at the non-silent codon positions.[10–11] Consequently, we will discuss the possible relationship between biased codon usage and the amino acid composition of proteins. We conclude that the interaction of biased DNA repair and mutation may influence, not only the difference in codon bias between species, but also the observed differences between genes within a single genome.

Gene Duplication and Gene Conversion

We will deal primarily with the phenomenon of codon bias in complex multicellular eukaryotes, particularly in Drosophila. Metazoan genomes are typically much larger than those of single celled microbes and many genes are present in multiple copies per haploid genome. A general property of gene families is that the family members can evolve in a concerted fashion.[12] Concerted evolution can result from a number of distinctly different molecular mechanisms, such as unequal crossing over or gene conversion. We will consider the case of gene conversion only. There is now ample evidence for the occurrence of gene conversion both during the course of evolution[13,14] and in laboratory experiments.[15–17] The experimental evidence points to a mechanism of gene conversion that parallels the stages of normal homologous recombination: sequence pairing, strand exchange, branch migration, heteroduplex formation and mismatch repair. Nucleotide mismatches between the duplicated coding sequences are "repaired" by the enzymes that are normally involved in other types of DNA repair within the cell. It is known that mismatch repair patterns are themselves highly biased,[17] and it is thought that this repair bias may be the result of natural selection to counter mutational biases. For instance, if mutation tends to increase the frequency of AT base pairs, then repair patterns will evolve to counteract this bias by favoring repair in the direction of GC pairs. This leads to the prediction that coding sequences that are subjected to the activity of these repair enzymes most frequently would be characterized by a high GC content that reflected this repair bias, rather than the normal level of GC that results from an equilibrium between mutational bias and the countervailing repair bias.

The potential for gene conversion and concerted evolution is highest among those genes that occur in linked clusters in the genome. The amylase genes of *D. melanogaster* provide an example of this phenomenon.[14] This gene pair undergoes concerted evolution due to gene conversion between the two coding sequences.[18] Sequence analysis confirmed the fact that the duplicated gene sequences were evolving as a pair, and also revealed that the GC content of the third codon position is almost 90%.[14] This is an unusually high GC content for Drosophila genes.[5] Not surprisingly, this high GC content in the silent positions is reflected in a very biased distribution of synonymous codons. It is probable that the two unusual features of the amylase gene pair, i.e., the high levels of gene conversion and the extreme bias in codon usage, have a common basis. Specifically, during the course of sequence homogenization, these sequences are exposed repeatedly to the effects of biased DNA repair. The resulting GC pressure would, of course, affect all nucleotide positions equally, but natural selection for protein function, acting at the non-synonymous sites, would tend to counteract this pressure. Thus, in these genes the extreme codon bias at the synonymous sites is not a reflection of natural selection for optimal gene function; rather, it is due to the relaxation of selection at the silent sites.

Gene Conversion and Codon Bias

In order to establish a firm relationship between the frequency of gene conversion and the degree of codon bias, one needs to have evidence from more than a single gene pair. Also, it would be useful to have a single gene family where some genes were undergoing concerted evolution while others were not. The trypsin gene family in *D. melanogaster* provides such an example. We have recently sequenced a 13kb region of DNA that includes the four known trypsin genes.[19] The sequencing revealed the presence of four additional trypsin coding sequences tightly clustered within this region. The set of eight genes includes two divergently-transcribed pairs, in addition to four other genes. The gene pairs are undergoing concerted evolution while the other genes have unique sequences that have diverged independently from the other family members. All eight genes encode functional trypsin proteins. Thus, the trypsin gene cluster provides an opportunity to monitor the effects of varying levels of gene conversion within a single gene cluster.

We examined the GC content and the synonymous codon distributions in these trypsin-encoding genes, and a sample of the results is presented in Table 1. The eta-trypsin sequence is highly diverged from the other Drosophila trypsins (more than 40%), while the gamma-trypsin is a member of a gene pair where there is less than 2% sequence divergence between the two members of the gene pair. A comparison of the two sequences should allow us to see the effects of gene conversion on both GC content and codon usage. The GC content of silent sites in the independently evolving eta-trypsin gene is 54% while in the gamma-trypsin it is 79%. This result indicates that the interconversion of the gamma-trypsin with its partner, the delta-trypsin gene, has resulted in an increase in the GC content at the silent sites. This elevated GC content is reflected in an increase in the number of codons ending in C (see Table 1). In the Table, only Serine and Threonine codons are shown to illustrate the general trends. It is obvious that all codons are used in significant frequencies in the eta-trypsin gene but that there is a bias towards codons ending in C in the gamma-trypsin sequence. One is left with the question of whether the bias towards NNC codons is a cause or

Table 1. Distribution of Serine and Threonine codons in two Drosophila trypsin genes.

	Serine			Threonine	
	Eta	Gamma		Eta	Gamma
TCT	3	4	ACT	2	1
TCC	7	25	ACC	3	10
TCA	3	0	ACA	3	0
TCG	4	0	ACG	2	0
AGT	4	1	—	—	—
AGC	5	15	—	—	—

One of these genes (gamma-trypsin) is subject to gene conversion while the other (eta-trypsin) is not. The numbers of each codon that occur within the coding sequence of each gene are shown. The overall GC content in the third codon position of the Eta trypsin is 54% while, for the gamma trypsin it is 79% (Wang et al., in prep).

an effect of the elevated GC content at the third codon positions. Some features of the data may help us to decide between these alternatives. First, for the two four fold degenerate codons shown, the most frequent codons, TTC and ACC, both end in C. This is also true of all the other four fold degenerate codons. One might wonder why all abundant tRNAs happen to be those that match codons ending in C. We have chosen the Serine codons to illustrate a second, related point. Serine has a total of six condons and we see from Table 1 that all six are used in the eta-trypsin gene. Two of the six, TCC and AGC, are used predominantly in the gamma-trypsin sequence. This means that there are two "preferred" codons for Serine, both of which end in C. It is difficult to see how selection for an optimal condon for Serine could select two codons, both of which happen to end in a C. This observation can be generalized to other genes and species (see Table 2). The most parsimonious explanation is that some evolutionary force, other than selection based on the distribution of individual tRNA abundances, is increasing the frequency of C in the third position of these codons. The correlation between gene conversion and degree of bias suggests that this force is biased DNA repair.

Codon Bias and Amino Acid Substitution

The extreme codon bias in genes undergoing concerted evolution is consistent with the view that the codon bias is, itself, merely a reflection of an underlying evolutionary pressure that affects the entire DNA sequence. Negative selection at the amino acid sequence level explains why such evolutionary pressures are not so obvious at the non-synonymous positions. Amino acid sequences can change over time, however, and if an evolutionary bias at the DNA level exists, its effects should eventually be seen at the non-synonymous positions as well. In order to look for such an effect, we need to compare homologous genes in distantly-related organisms that are characterized by a wide range of GC contents.

Table 2. Distribution of Serine codons in the alpha-amylase genes of Tribolium, Drosophila and Streptomyces.

	Tribolium	Drosophila	Streptomyces
TCT	3	0	0
TCC	2	26	19
TCA	5	0	0
TCG	5	4	4
AGT	7	1	0
AGC	6	18	11

The GC contents at the 3rd position of codons in the Tribolium,[31] Drosophila[32] and Streptomyces[33] sequences are 49%, 89%, and 95%, respectively. The numbers refer to the occurrence of each codon within each gene.

In this case, we are not focusing on the causes of the variance in GC content, but rather on its possible relationship with the amino acid content of the encoded protein. For instance, we do not know if the extreme GC bias in the *Streptomyces limosus* sequence (see Table 2) is due to mutational bias or to a repair bias.

The distribution of Serine codons in the alpha-amylase genes of two insects and one prokaryote are shown in Table 2. The two insect sequences show more than 60% sequence identity at the amino acid level but have very different GC contents in the third position of codons. The prokaryotic amylase has lower, but very significant, levels of amino acid sequence similarity with the two insect sequences (approximately 44%) and is characterized by a very high GC content at the third codon position. As can be seen from the Table, the Tribolium sequence contains all six Serine codons in approximately equal frequencies. What is remarkable is the fact that organisms as different as Drosophila and Streptomyces show strikingly similar biases in the distribution of their Serine codons, including a preference for both NNC codons. Again, the data for Serine codons are shown as an example of what is also true for other codon groups in these genes.

A comparison of amino acid frequencies in these three genes is shown in Table 3. Our expectation is that genes that are biased towards a high GC content will gradually accumulate amino acids whose codons are relatively GC-rich. We find that this is, indeed, the case. The frequency of Arginine and Alanine codons, both of which are relatively GC-rich show a positive correlation with the TC content at silent sites, while the relatively AT-rich codons for Isoleucine, Asparagine and Phenylalanine show a negative correlation. There is no reason that such correlations should exist if the GC content at the silent sites was due primarily to variation in the abundance of tRNAs.

The data shown in Tables 2 and 3 give an indication of the differences in response time between silent sites and non-silent sites to an overall change in nucleotide composition. In Table 2, we see that the Drosophila and Streptomyces genes resemble one another in their degree of codon bias, and are in contrast to the Tribolium gene. In Table 3, however, the biggest differences are between

Table 3. Relationship between GC content at silent sites and variation in the amino acid composition of alpha-amylase genes.

		Tribolium	Drosophila	Streptomyces
GC (3rd position):		49	89	95
GC-rich codons:	Arg	17	21	32
	Ala	41	41	65
AT-rich codons:	Ile	30	20	11
	Asn	40	35	19
	Phe	23	23	16

The GC content at the third codon position is expressed as a percentage. The absolute number of codons for each amino acid is shown (includes all synonymous codons).

the Tribolium and Streptomyces genes; the Drosophila amylase is intermediate, but closer to the Tribolium gene. These patterns are consistent with the fact that the Drosophila amylase has been under GC pressure only since the gene duplication event, whereas the Streptomyces amylase is found in a very GC-rich genome[20] which has presumably been under GC-pressure for a much longer time. In other words, the Drosophila amylase gene has responded to the nucleotide pressure at the non-selected silent sites, but is only beginning to respond at the non-silent sites.

Discussion

Many of the points made above have already been discussed in the literature. For instance, the relationship between the GC content of silent sites and codon bias is well recognized.[8] The influence of GC content on amino acid composition has also been evaluated by comparing different organisms[7,21] and by comparing many genes within a single organism.[11,22] Even our suggestion that high codon bias might be related to concerted evolution has been noted previously.[23] What we have tried to add to this debate is direct evidence that the correlations seen in whole genome comparisons can also be seen in phylogenetic comparisons of individual genes; we also provide an explicit mechanistic explanation of why duplicated genes should show high codon bias.

Ikemura[1] concludes that while codon bias in unicellular prokaryotes may have a selective basis, the same phenomenon in multicellular eukaryotes may be due to other forces such as mutational bias. This same view is echoed in the study of mammalian genes by Wolfe et al.[8] Among higher eukaryotes, the best evidence for natural selection acting on codon bias comes from the study of *D. melanogaster* genes.[5] This evidence is based on the argument that the observed heterogeneity between Drosophila genes, coupled with the clear difference in GC content between coding and non-coding sequences, cannot be interpreted simply as the result of a common mutational bias. We suggest that variation in the levels of biased recombinational repair within gene families can provide a non-selective explanation for the heterogeneity in GC content between genes within the Drosophila genome. Not only does gene conversion provide an explanation for why some genes are GC-rich, it also explains why the adjacent non-coding sequences are not also GC-rich. This is because the effects of sequence homogenization end abruptly at the end of the coding sequences; this is true for both the amylase gene pair[14] and the trypsin gene family (Magoulas and Hickey, submitted).

Recently, Kliman and Hey[24] reported that there is a negative correlation between the degree of codon bias and recombination frequency in Drosophila. They interpret this result as being indicative of reduced selection for optimal codons in regions of low recombination. Alternatively, these data might be interpreted as indicative of the increased GC pressure caused by elevated rates of recombinational repair.

In higher eukaryotes, most instances of codon bias involve an increase in the frequency of codons ending in C or G. A notable exception is the mitochondrial genome, where the bias is in favor of A and T.[25] This dramatic difference between mitochondrial genes and nuclear genes, even within a single species such as *D. melanogaster,* actually lends support to our arguments. Animal mitochondria lack DNA repair mechanisms: consequently, their sequences reflect the pressures of mutation bias, but not repair bias. Furthermore, organelle coding sequences are relatively rich in amino acids such as Isoleucine and poor in Alanine.[20,26] Again, this supports the general correlation between nucleotide bias, codon bias and amino acid composition. In fact, mitochondrial genomes illustrate a further level of bias. In addition to adjusting synonymous codon frequencies and amino acid composition, some organelle genomes have modified their genetic code in order to better cope with the extreme nucleotide bias.[27,28]

Much of the difficulty in interpreting the causes of codon bias stems from the fact that a high level of GC bias (or AT bias) automatically results in codon bias). The converse is not true, i.e., one could have a nonrandom distribution of synonymous codons without altering the overall GC content. The fact that in many cases all of the codons are biased in favor a single nucleotide is also the cause of a potential problem in phylogenetic analyses. Namely, DNA sequences that share the same GC bias might cluster more closely than if such a bias did not occur. This effect has recently been discussed in relation to phylogenetic trees based on rRNA sequences.[29] It is usually assumed that the use of inferred amino acid sequences would allow us to avoid such biases in the case of protein-coding genes. The observation that amino acid composition correlates with GC bias indicates that even amino acid sequences are not immune to such biases. In practice, this would only become a problem when one is using protein-coding sequences to infer very ancient divergences.[30]

What we have presented in this paper is essentially a "neutralist" interpretation of codon bias. The distribution of codons in GC-rich genes is the result of a balance between the GC pressure of DNA repair which tends to eliminate certain codons, and the effects of mutation which regenerates these lost codons. The correlations with amino acid content point to the fact that there is a similar balance being struck at some of the non-silent positions. The emerging body of population-level sequence data will provide a test for these ideas. In particular, we expect the distribution of nucleotide frequencies at polymorphic sites to match the distribution of synonymous codons. In other words, at any given polymorphic site in a Drosophila gene that shows high codon bias, the most common variant would be expected to be a C. This should be true, not only for silent polymorphisms, but also for non-silent, non-selected, polymorphisms.

Conclusion

In summary, we point out that gene duplication predisposes coding sequences to gene conversion. Gene conversion involves biased heteroduplex repair which can lead, in turn, to an elevated GC content in the duplicated genes. This GC pressure would affect all nucleotides equally, but is counteracted by selection at non-synonymous sites. Consequently, synonymous sites evolve to give a biased distribution of codons. Even at the replacement sites, however, this GC pressure can be seen in the longer term. Both the rapid response to GC pressure at the silent sites, and the slower response at the non-silent sites, reflect a common evolutionary mechanism. This explanation of biased codon usage in terms of biased heteroduplex repair may apply to only a small fraction of genes. It does, however, illustrate the general principal that molecular events at the DNA level can influence both the choice of synonymous codons and, to a lesser extent, the amino acid composition of proteins. What is perhaps of most interest is that it provides an example where clear deviations from the expectations of a simple neutral mutation model do not, automatically, point to natural selection acting at either the level of translational efficiency or amino acid sequence.

Acknowledgments

This research was supported by grants from NSERC Canada and MRC Canada, and by a Fellowship from the Canadian Institute for Advanced Research (D.A.H.)

References

1. T. Ikemura. (1985). *Mol. Biol. Evol.* 2, 13–34.
2. P. M. Sharp and W-H. Li. (1986). *J. Mol. Evol.* 24, 28–38.
3. W-H. Li. (1987). *J. Mol. Evol.* 24, 337–345.
4. M. Bulmer. (1987). *Nature* 325, 728–730.
5. D. C. Shields, P. M. Sharp, D. G. Higgins, and F. Wright. (1978). *Mol. Biol. Evol.* 5, 704–716.
6. E. N. Moriyama and D. L. Hartl. (1993). *Genetics* 134, 847–858.
7. N. Sueoka. (1988). *Proc. Natl. Acad. Sci.* (USA) 85, 2653–2657.
8. K. H. Wolfe, P. M. Sharp, and W-H Li. (1989). *Nature* 337, 283–285.
9. J. P. Carulli, D. E. Krane, D. L. Hartl, and H. Ochman. (1993). *Genetics* 134, 837–845.
10. N. Sueoka. (1993). *J. Mol. Evol.* 37, 137–153.
11. D. W. Collins, and J. H. Jukes. (1993). 36, 201–213.
12. N. Arnheim, M. Krystal, R. Schnickel, G. Wilson, O. Ryder, and E. Zimmer. (1980). *Proc. Natl. Acad. Sci. (USA)* 77, 7323–7327.
13. Y. Xiong, B. Sakaguchi, and T. H. Eickbush. (1988). *Genetics* 120, 221–231.
14. D. A. Hickey, L. Bally-Cuif, S. Abukashawa, V. Payant, and B. F. Benkel. (1991). *Proc. Natl. Acad. Sci. (USA)* 88, 1611–1615.
15. K. K. Willis, and H. L. Klein. (1987). *Genetics* 117, 633–643.

16. A. Letsou and R. M. Liskay. *Genetics* 117, 759–769.

17. T. C. Brown, and J. Jiricny. (1988). *Cell* 54, 705–711.

18. V. Payant, S. Abukashawa, M. Sasseville, B. F. Benkel, D. A. Hickey, and J. David. (1988). *Mol. Biol. Evol.* 5, 560–567.

19. C. A. Davis, D. C. Riddell, M. J. Higgins, J. J. A. Holden, and B. N. White. (1985). *Nucl. Acids Res.* 13, 6605–6619.

20. K. Wada, Y. Wada, H. Doi, I. Ishibashi, T. Gojobori, and T. Ikemura. (1991). *Nucl. Acids Res.* 19 (Suppl), 1981–1985.

21. N. Sueoka. (1961). *Proc. Natl. Acad. Sci (USA)* 47, 1141–1149.

22. G. D'Onofrio, D. Mouchiroud, B. Aissani, C. Gauthier, and G. Bernardi. (1991). *J. Mol. Evol.* 32, 504–510.

23. P. M. Sharp. (1991). *J. Mol. Evol.* 33, 23–33.

24. R. M. Kliman and J. Hey. (1993). *Mol. Biol. Evol.* (in press).

25. D. R. Wolstenholme and K. W. Jeon. (1992). *Mitochondrial Genomes,* Academic Press.

26. S. T. Aota, T. Gojobori, F. Ishibashi, T. Maruyama, and T. Ikemura. (1988). *Nucl. Acids Res.* 16, 315–402.

27. S. G. E. Andersson and C. G. Kurland. (1991). *Mol. Biol. Evol.* 8, 530–544.

28. S. Osawa, D. Collins, T. Ohama, T. Jukes, and K. Watanabe. (1990). *J. Mol. Evol.* 30, 322–328.

29. M. Hasegawa and T. Hashimoto, *Nature* 361, 23.

30. M. A. Steel, P. J. Lockhart, and D. Penny. (1993). *Nature* 364, 440–442.

31. D. A. Hickey, B. F. Benkel, P. H. Boer, Y. Genest, S. Abukashawa, and G. Ben-David. (1987). *J. Mol. Evol.* 26, 252–256.

32. P. H. Boer and D. A. Hickey. (1986). *Nucl. Acids Res.* 14, 8399–8411.

33. C. M. Long, M-J. Virolle, S-Y. Chang, S. Chang, and M. J. Bibb. (1987). *J. Bacteriol.* 169, 5745–5754.

17

Genealogical Portraits of Speciation in the *Drosophila melanogaster* Species Complex

Jody Hey and Richard M. Kliman

Introduction

A particular class of DNA sequence data set, one that is increasingly being generated for purposes of exploring natural selection, can also be used to study speciation. Minimally, these analyses require, for each of multiple loci, several DNA sequences from each of the species under investigation. The analysis of a recently generated data set for the four species of the *Drosophila melanogaster* complex is reviewed within the context of speciation. For three loci, a consistent pattern of variation emerges, in which it appears that ancestral *D. simulans* gave rise to both *D. mauritiana* and *D. sechellia*. Surprisingly, modern *D. simulans* appears to retain much of the genetic variation that existed prior to those speciation events.

This report describes recent progress in the application of modern population genetic methods to speciation problems. The basic methodology draws heavily from advances in genealogical modeling and from the use of DNA sequence data for studying natural selection. In recent years, the theoretical arm of population genetics has seized onto genealogical models (also called coalescent models). The paradigm has shifted away from a focus on allele frequencies and toward a focus on the structure of gene trees. Much of this shift has been driven by the actual data, simply because homologous DNA sequences are readily interpreted in terms of estimates of historical gene trees. As the nature of the data has changed, so have the models.[1,2]

The synergism of theoretical and empirical research has been especially effective in the investigation of natural selection acting on genetic loci, and several recent reports present statistical evidence for adaptive natural selection at or near the loci studied. The cases include fairly recent occurrences of balancing selection[3] and strong directional selection.[49] The methods used in these studies of natural selection draw on two types of contrasts: (1) comparison of variation within species to that between species; and (2) comparison of variation among genomic regions (sometimes different loci and sometimes different portions of a single locus). These contrasts take place within the framework of a null model, in which each species has a constant population size and is not subject to natural selection. The model permits variation in the neutral mutation rate among loci (but not among species for a given locus), and also permits variation in population size among species (but not among loci for a given species). In short, the model assumes that the ratio of within-species variation to between-species variation will be the same for all of the loci. The resulting statistical test can take the form of a simple chi-square design,[6] or it can require more complicated strategies.[3]

The approach of using multiple genomic regions and multiple species lends itself to the study of speciation. Because the null model assumes that the relationship between intraspecific and interspecific variation is the same for all regions of the genome, the model can be used to test whether or not all loci are consistent in terms of divergence during speciation. In other words, these methods can be used to statistically test whether the pattern of variation among a set of loci from different species is consistent with a model in which the timing of the cessation of gene flow during species formation is the same for all of the loci.

A Case Study

The remainder of this review summarizes recent findings from a study of the four species of the *Drosophila melanogaster* species complex and of three X-linked loci[10,11]: approximately 1900 base pairs of the *period* locus (*per*), 1000 base pairs of the *zeste* locus, and 1100 base pairs of the *yolk protein* 2 locus (*yp*2). Two of the species, *D. melanogaster* and *D. simulans,* are cosmopolitan, while the other two, *D. mauritiana* and *D. sechellia,* are endemic to oceanic islands.[12] Numerous phylogenetic and species-hybridization studies have revealed only that *D. melanogaster* is a sister-taxon to the other species. *Drosophila simulans, D. sechellia* and *D. mauritiana* are morphologically very similar to each other and, despite considerable effort, a bifurcating phylogeny for these species has not been clearly resolved.[13–16] Collectively, these four species represent three speciation events, two of which appear to have occurred very recently and at similar times.

X-linked loci were chosen for the simple convenience that DNA preparations from individual male flies avoids heterozygosity in PCR-generated DNA sequences. Each species and locus was represented by six isofemale lines (*i.e.*, the laboratory strain was started with a single wild caught female) collected from multiple locations in the species range (excepting *D. mauritiana,* which is found only on the island of Mauritius). One DNA sequence was generated from each line for each locus. With the exception of five of the *zeste* gene sequences, all three genes were sequenced from the same DNA preparations, meaning that the three sequences for each strain actually came from the same chromosome. In the case of the *zeste* exceptions, the sequence still came from the same strains.

Figure 1 shows estimates of the amount of variation within each species for each locus. Figure 2 shows the divergence between *D. melanogaster* and *D. simulans,* per base pair, for each locus. Both *per* and *zeste* appear more variable than does *yp*2, and this pattern appears both in the intraspecific comparisons (Figure 1) and the interspecific comparisons (Figure 2). *Drosophila sechellia* has less variation than the other species, and this pattern is consistent among loci. Despite considerable variation among loci and among species, a casual appraisal suggests that this variation is consistent with the neutral model. The fit of the null model can be tested more formally using the procedure of Hudson, Kreitman and Aguadé[3] (the procedure is now widely used and frequently referred to as the HKA test). When this method was applied to the data, the overall fit of the model was quite good. The method also returns scalars of proportionality for population sizes and mutation rates. Relative to *D. melanogaster* (1.0), the population sizes for the other species are 1.600 for *D. simulans,* 1.296 for *D. mauritiana,* and 0.111 for *D. sechellia*. Similarly, for mutation rates, the scalars of proportionality relative to *zeste* are 2.372 for *per* and 0.722 for *yp2*. In sum, there is no evidence for recent directional or balancing selection; and the data are consistent with a neutral model in which population sizes vary across species,

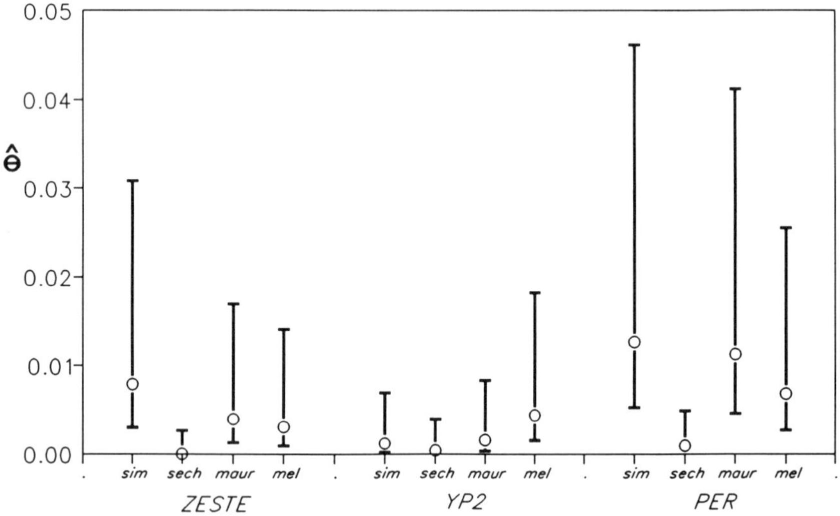

Figure 1. Estimates of $4Nu$ ($\hat{\theta}$) per base pair, where N is the effective population size of a species and u is the neutral mutation rate. $\hat{\theta}$, and the 95% confidence intervals were calculated from the number of polymorphic sites observed in each species and locus according to the procedure of Kreitman and Hudson.[27]

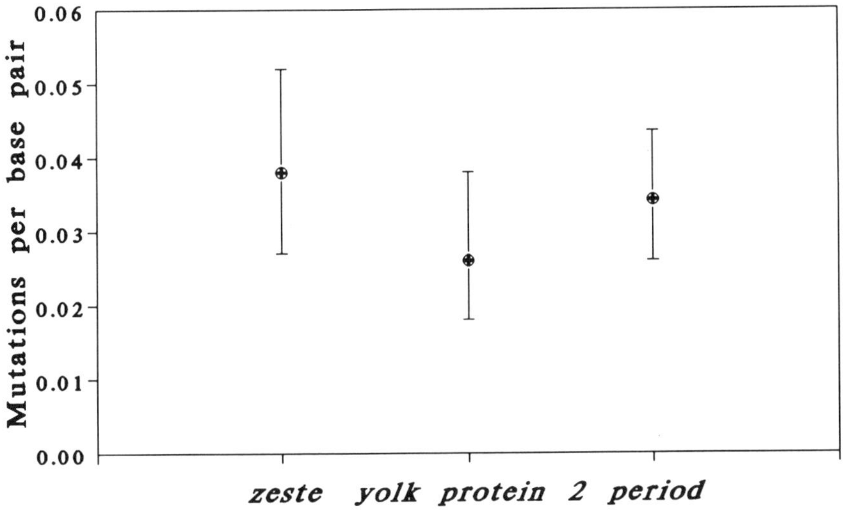

Figure 2. The average number of differences per base pair separating *D. melanogaster* sequences from *D. simulans* sequences. The 95% confidence intervals where generated assuming a Poisson distribution of differences.

but not across loci, and in which neutral mutation rates vary across loci, but not species.

Figure 3 shows tree diagrams generated with the neighbor-joining procedure.[17] The diagrams can be considered in two different ways. They are first, and most literally, the output of a particular clustering algorithm that joins sequences and groups of sequences by a complicated assessment of sequence similarity and tree length. Secondly, these diagrams may be considered as estimates of the historical genealogy of the gene copies in the sample. In this light, the tips of branches refer to the actual DNA sequences, and the branches refer to the persistence of ancestral DNA sequence lineages through time. The nodes of the tree refer precisely to DNA replication events in which both copies produced are ancestors of sequences included in the sample.

One way a branching tree diagram may not fit a genealogy is if the history includes recombination among the ancestral sequences of a sample.[18] In fact, the *per* locus data suggest a history with considerable recombination.[10] The patterns of variation, especially within *D. simulans* and *D. mauritiana,* suggest that a bifurcating diagram like that in Figure 3 is not a good model of the genealogy. Because of recombination, some parts of the *per* gene in these species have different genealogical histories than do other parts of the gene. Other evidence of recombination at *per* in these species is the observation that 11 polymorphisms are shared between them. Even though the neighbor-joining algorithm separated all the *D. mauritiana* sequences from the *D. simulans* se-

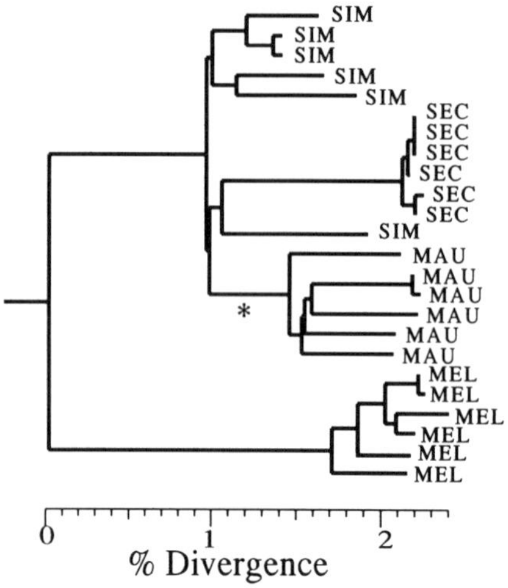

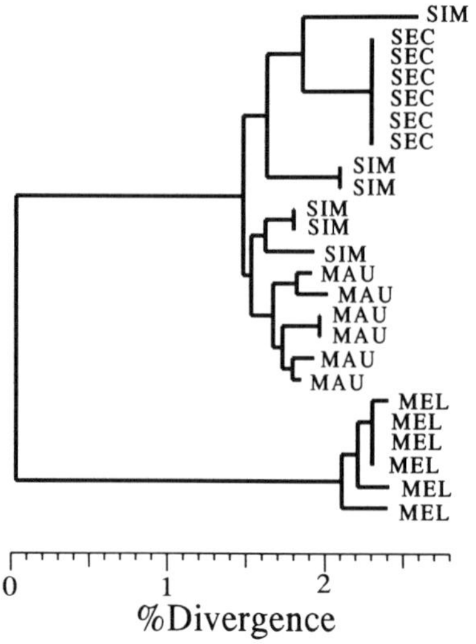

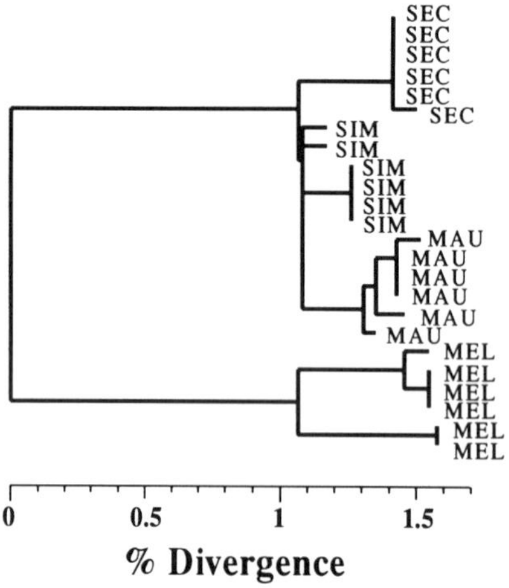

Figure 3. Neighbor-joining distance trees. **A,** the *period* locus (redrawn from figure 6 of Kliman and Hey[10]). The (*) marks the branch separating *D. mauritiana* sequences from those of other species—see text. **B,** the *zeste* locus (redrawn from figure 5 of Hey and Kliman[11]). **C,** the *yolk protein* 2 locus (redrawn from figure 5 of Hey and Kliman[11]).

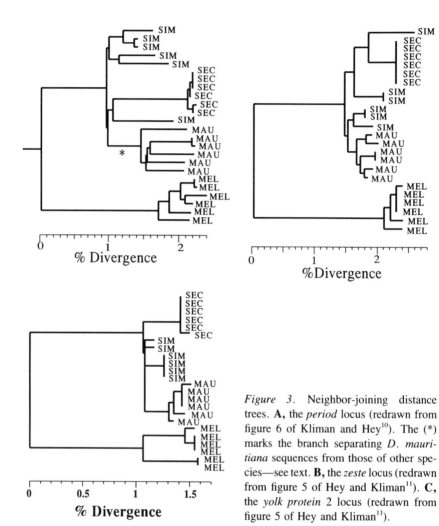

quences, this part of the diagram cannot be interpreted as a genealogy. The branch of the tree marked with an asterisk (Figure 3A) cannot represent a historical DNA sequence, for the simple reason that polymorphisms cannot be carried by a single DNA sequence.

Recombination at *per* notwithstanding, the three trees share several characteristics: *D. melanogaster* lineages are clearly separated from those of the other species, consistent with all other phylogenetic studies on this group[12]; all of the lineages within *D. schellia* and *D. mauritiana* form discrete clusters; and the earliest nodes of the tree, excluding the split between *D. melanogaster* and the other species, separate lineages of *D. simulans*.

Considering Speciation

Because no evidence was found for recent balancing or directional selection at these loci, and since they have similar genealogical histories, the data can be interpreted in terms of evolutionary forces that are expected to affect all loci in the same way. In other words, with no evidence that locus-specific forces have been at work, the data are more useful for considering evolutionary forces that affect all loci in the same way.

The shapes of the gene trees show that present day *D. simulans* still has genetic variation that predates the origin of the island endemic species. All three of the gene trees show *D. simulans* lineages extending to nodes that occurred prior to the base nodes of the *D. mauritiana* and *D. sechellia* clusters. It appears that *D. mauritiana* and *D. schellia* arose independently from ancestral *D. simulans,* and that modern *D. simulans* has changed relatively little since that time. Since modern day *D. mauritiana* and *D. sechellia* are endemic to different oceanic islands, a plausible model for their formation is that they diverged from ancestral *D. simulans* following the isolation of a small number of individuals that successfully crossed the oceanic barrier. It has been suggested that this kind of "founder" event, involving only a small number of individuals, describes a possibly frequent scenario for speciation.[19–21] However, this does not appear to be the case for *D. mauritiana,* which has nearly as much variation as does *D. simulans,* indicating a large effective population size. Also, 11 of the *per* polymorphisms are shared between *D. mauritiana* and *D. simulans;* it is unlikely that such a number of presumably ancient polymorphisms would persist if *D. mauritiana* ever experienced a population bottleneck. Thus, a very small effective population size for *D. mauritiana* at any time in its history seems unlikely. The repeated observation of very little variation in *D. sechellia* indicates a small effective population size for this species, which is consistent with speciation via "founder" event. However, one can only say that *D. sechellia* has had a small population size recently, and it remains possible that the population size was larger in the past.

Linking a Gene Tree Approach with a Gene Mapping Approach

The *zeste, yp*2, and *per* data reveal a unique view of population sizes during and since species formation. However, by themselves, these data do not inform on the role of genetic variation in the process of speciation. Consider loci that were in some fashion affected by natural selection as a component of the speciation process. One can imagine a history of differential adaptation, with different functional alleles suited to different environments; alternatively, there may be underdominance at some loci whereby heterozygotes (*i.e.*, hybrids) are less fit. Regardless, there are a variety of speciation models in which, because of natural selection, some loci experience different patterns of gene flow between incipient species than others. If there is opportunity for some gene flow while reproductive isolation develops, those loci that are not included in the gene flow will have different gene trees than those that are included.

The *zeste, yp*2 and *per* data provide an excellent opportunity for testing whether specific loci have played an active role in species formation. If candidates for such "speciation genes" are found, and if sequence data is obtained from within and between the several species, then the contrasting patterns between loci may well inform on the historical role of such "speciation genes." In short *per, zeste* and *yp*2 may serve as reference loci.

One way to include loci associated with speciation is to map, at a very fine scale, those loci responsible for fitness loss in hybrids. Once identified, a locus can be examined for intraspecific and interspecific variation in a manner parallel to that for *zeste, yp*2 and *per*. Not all of these loci, and possibly none, have necessarily contributed to speciation, as hybrid fitness loss is expected to accumulate after speciation. However, for very recent speciation events, there is the chance that loci identified in this way may inform on speciation. Significant progress with this approach has recently been made using *D. simulans-D. mauritiana* hybrids and *D. simulans-D. sechellia* hybrids.[22,23] In particular, a small gene region, containing a putative locus named *Odysseus,* has been located that appears to contribute to reproductive isolation between *D. simulans* and *D. mauritiana*.[23] Interestingly, this gene does not seem to contribute to reproductive isolation between *D. simulans* and *D. sechellia,* consistent with independent evolution of reproductive isolation from *D. simulans* in the two island species.

In some cases, it may also be possible to map loci responsible for premating isolation, though quantitative analysis of behavioral traits can be difficult. In the *D. simulans* complex, at least two loci appear to contribute to premating isolation between *D. sechellia* females and *D. simulans* males, while at least three loci (including two on the second chromosome) contribute to sexual isolation between *D. mauritiana* females and *D. simulans* males.[24,25] It is unlikely that this represents a single, shared evolutionary process (*i.e.,* that the two island species formed from a single species that evolved premating isolation from *D. simulans*). First, *D. sechellia* and *D. mauritiana* are, themselves, reproductively isolated. Second, the quantitative effects on sexual isolation of loci linked to specific autosomes differs in the two species, as does the degree of dominance. Third, the behavioral bases to sexual isolation differ, with *D. simulans* males actively courting *D. mauritiana* females, while avoiding *D. sechellia* females.[24] It should be added that, although courtship between *D. mauritiana* and *D. simulans* is often successful, the duration of copulation is short, resulting in the actual reproductive isolation.[25] These results are consistent with independent speciation events for the two island endemics, as also indicated by the molecular data.

The *zeste, yp*2 and *per* studies have laid a baseline that can be used to test the role of other loci in speciation. As mapping studies proceed, and other methods[26] for identifying candidate loci for involvement in speciation are developed, we may find a wider diversity of genealogical histories among loci.

Acknowledgments

This work was supported by National Science Foundation grant BSR 8918164.

References

1. W. J. Ewens. (1990). in *Mathematical and Statistical Developments of Evolutionary Theory* (S. Lessard, ed.) pp. 177–227, Kluwer Academic Publishers.
2. R. R. Hudson. (1990). In *Oxford Surveys in Evolutionary Biology* (*Vol. 7*) (P. H. Harvey and L. Partridge, ed) pp. 1–44 Oxford University Press.
3. R. R. Hudson, M. Kreitman, and M. Aguadé. (1987). *Genetics* 116, 153–159.
4. A. J. Berry, J. W. Ajioka, and M. Kreitman. (1991). *Genetics* 129, 1111–1117.
5. D. Begun and C. F. Aquadro. (1991). *Genetics* 129, 1147–1158.
6. J. H. McDonald and M. Kreitman. (1991). *Nature* 351, 652–654.
7. W. Stephan and S. J. Mitchell. (1992). *Genetics* 132, 1039–1045.
8. C. Langley, J. M. MacDonald, N. Miyashita, and M. Aguadé. (1993). *Proc Nat. Acad. Sci. USA* 90, 1800–1803.
9. W. F. Eanes, M. Krichner, J. Yoon. (1993). *Proc. Natl. Acad. Sci. USA* 90, 7475–7479.
10. R. M. Kliman, and J. Hey. (1993). *Genetics* 133, 375–387.
11. J. Hey and R. M. Kliman. (1993). *Mol. Biol. Evol.* 10, 804–822.
12. D. Lachaise, M.-L. Cariou, J. R. David, F. Lemeunier, L. Tsacas, and M. Ashburner. 1988. *Evolutionary Biology* 22, 159–225.
13. M. Bodmer and M. Ashburner. (1984). *Nature* 309, 425–430.
14. V. H. Cohn, M. A. Thompson, and G. P. Moore. (1984). *J. Mol. Evol.* 20, 31–37.
15. J. A. Coyne and M. Kreitman. (1986). *Evolution* 40, 673–691.
16. A. Caccone, G. D. Amato, and J. R. Powell. (1988). *Genetics* 118, 671–683.
17. N. Saitou and M. Nei. (1987). *Mol. Biol. Evol.* 4, 406–425.
18. R. R. Hudson and N. L. Kaplan. (1988). *Genetics* 120, 831–840.
19. E. Mayr. (1954). In *Evolution as a Process* (J. Huxley, C. Hardy, and E. B. Ford, eds.) pp. 157–180, Allen & Unwin.
20. H. L. Carson. (1975). *Am. Nat.* 109, 83–92.
21. A. R. Templeton. (1980). *Genetics* 94, 1011–1038.
22. C.-I. Wu, D. E. Perez, A. W. Davis, N. A. Johnson, E. L. Cabot, M. F. Palopolis,

and M.-L. Wu. (1993). In *Mechanisms of Molecular Evolution* (N. Takahata and A. G. Clark, ed.) pp. 191–212, Sinaur.

23. D. E. Perez, C.-I. Wu, N. A. Johnson, and M.-L. Wu. (1993). *Genetics* 134, 261–275.
24. J. A. Coyne. (1992). *Genet. Res., Cambr.* 60, 25–31.
25. J. A. Coyne. (1993). *Evolution* 47, 778–788.
26. L.-W. Zeng. and R. S. Singh. (1993). *Genetics* 135, 135–147.
27. M. Kreitman and R. R. Hudson. 1991. *Genetics* 127, 565–582.

18

Genetic Divergence, Reproductive Isolation and Speciation

Rama S. Singh and Ling-Wen Zeng

Discontinuities observed between species must have owed its origin to discontinuities occurring in the evolution of each. . . .

Bateson (1894)

Species and higher categories originate in single macroevolutionary steps as completely new genetic systems. The general process which is involved consists of a repatterning of the chromosomes, which results in new genetic system.

Goldschmidt (1940)

Introduction

Species are the most readily recognizable units in the diversity of life and the mechanism of speciation has always been and still is a central problem in evolutionary biology.[33] Almost all species with which we come in contact in daily life and certainly a large proportion of the formally described species show large gaps of qualitative or quantitative nature and this large gap has been a stumbling block to the study of speciation. This is for two reasons. First, presence of large gaps between species meant that species-specific traits could not be subjected to Mendelian genetic analysis, and second, the large gaps observed between species were used to propose theories of speciation which went against the neo-Darwinian mechanisms of gradual evolution (e.g., see Bateson 1894, Goldschmidt 1940). It is therefore not surprising that most genetic theories of speciation advocating macroevolutionary mechanisms of speciation, prior to the advent of molecular techniques in 1960s, were based on large changes in the genome such as chromosomal changes or macromutations.[23,59] On the other hand, the neo-Darwinian theories were based on a collection of genetic and ecological factors, with strong emphasis on natural selection and geographic isolation which was required to complete the job of reproductive isolation.[20,27,32,55]

The recognition that evolutionary processes of gradual evolution initially must produce species that are far from having complete reproductive isolation led to the discovery of the so called sibling species which are phenotypically similar but with complete or incomplete reproductive isolation. The sibling species with incomplete reproductive isolation have become the most suitable material for the genetic studies of pre-mating and post-mating reproductive isolation.

All problems of speciation are ultimately connected, directly, or indirectly, with the nature and the causes of genetic changes that occur before, during and after speciation (Table 1). Since systematic genetic studies of speciation could not be done before the molecular techniques became available, it is not surprising that all debates on speciation prior to the 1960s centered on sudden vs. gradual speciation (pre-1940) or on sympatric vs. allopatric speciation (1940–1960). It is only after the 1960s with the advent of gel electrophoresis that it became possible to mount a successful attack on the genetic basis of species differences and speciation. With molecular techniques we can ask the genetic questions that need to be asked, such as, the nature and the number of genes involved in reproductive isolation.

There have been basically two approaches to estimate the number of genes involved in speciation: the molecular approach and the classical genetic approach. In the first approach attempts have been made to estimate the total number of gene differences between related species (for recent reviews see Singh[51,52]). The second approach involves estimating number of genes or chromosomal factors affecting reproductive isolation by the use of phenotypic markers in the classical backcross studies.[13,20] The genetic and molecular analysis of post-zygotic reproductive isolation has become an interesting topic in evolutionary biology. Studies of this type are starting to shed light on not only the number of genes but also the nature of genes and genic interactions involved in post-zygotic reproductive

Table 1. The problems of speciation

1. The species concept
   - Sexual vs. asexual organisms
   - Population genetics vs. systematics
2. Geographic models of speciation
   - Allopatric
   - Sympatric
3. Genetic mechanisms of speciation
   - Genic
   - Chromosomal (incompatibility)
   - Genomic (major mutations)
   - Molecular "parasites"
4. Founder effects and speciation
   - Open/closed genetic systems
   - Genetic transilience
5. Major morphological and physiological changes
   - Coincide with speciation (macroevolution)
   - Follow speciation (microevolution)
6. Is speciation adaptive or non-adaptive?

isolation. The objectives of this paper are as follows: (1) to summarize data on the nature of genic differences within and between species and their bearing on the causes of reproductive isolation. We will show that the patterns of molecular variation within and between species are only loosely coupled and they are not related to the nature or magnitude of genetic differences causing reproductive isolation between species; (2) to summarize genetic studies of post-zygotic reproductive isolation with emphasis on the number and nature of genes and the nature of incompatible genic interactions involved in hybrid sterility and inviability that occur during the early stages of speciation, and (3) in view of the nature of genetic differences between species and in view of the number and nature of genes known to affect reproductive isolation, to briefly out line a *unified* genetic theory of speciation (Singh 1989, 1990) which will be fully developed elsewhere.

The Nature of Genetic Divergence Between Species

The largest proportion of observed genetic divergence between closely related species occurs at the level of nucleotides and amino acids.[22,31,34] The great majority of the data on genic difference between species comes from the application of gel electrophoresis on protein variation.[1,31,34,52] The body of DNA sequence divergence data on closely related species is still too sparse to enable us to say anything definitely about the genetic basis of speciation although such data are beginning to shed new lights on some old topics such as species relationships and some new topics such as the relationship between genic divergence and rates of recombination.[3,30,46,50] As the large body of the protein variation data has been summarized at numerous places (e.g., Lewontin,[31] Ayala,[1] Singh[51,52]) here we will highlight only the main conclusions. While we believe that these conclusions are supported by the data set from all organisms, they all in fact have emerged from the detailed analyses of *D. melanogaster* group species in our laboratory (Singh,[51] Choudhary, Coulthart and Singh,[7] Thomas and Singh,[57] Singh and Choudhary, unpublished results).

Lack of Correlation between Levels of Genic Divergence and Reproductive Isolation

The relationship between genic divergence and levels of reproductive isolation was the very first question that was settled by gel electrophoresis. The estimate of genic divergence between closely related species vary from zero or a few percent between partially isolated species or subspecies to about 10% between sibling species such as *D. melanogaster* and *D. simulans*.[1,7,56] Furthermore it is obvious that even the estimates between the most closely related species must necessarily be *overestimates* of the actual number of genes involved in speciation, as they contain both relevant and irrelevant changes with respect to speciation. The overall estimates of genic divergence between species are more a reflection of the time since divergence. However, the electrophroetic studies did not tell us what proportion of genes are involved in speciation, they did tell us that the proportion of such genes is definitely very small—too small to be detected by such studies!

Lack of Correlation between Geographic Differentiation (Fst) within Species and Genetic Distance (D) between Species

There is no predictable theoretical relationship between Fst and D as the former measures the cumulative effects of genetic drift and gene flow between populations, and the latter in a function of mutation rates and time since divergence between species. However we reasoned that a positive relationship between estimates of Fst and D for individual loci would be an indication of a gradual selective (or non-selective) divergence in large populations, and a negative relationship, of rapid fixation of alleles due to selection or population bottlenecks. *D. melanogaster* and *D. simulans* are a pair of sibling species with almost similar cosmopolitan distribution so that a proper comparison of their Fst and D can be made. We observed a negative but non-significant correlation between Fst and D which means that *loci with high Fst within species tend to be similarly monomorphic between species.*[7] We observed only three out of a total of 61 polymorphic loci which had both high Fst and high D, but two of these loci had high D because they had procured new alleles in one of the two species. Thus the negative relationship between Fst and D suggests that loci with high D have achieved their status (moved towards monomorphisms) either by bottleneck or by selection. In either case the behavior of polymorphic loci between these two species suggests that either one (probably *D. simulans*) or both of them have gone through bottleneck, or that they both have responded to different selection pressure. The latter scenario is not surprising as one can hardly believe that selection coefficient would remain constant over the life time of a species.[22]

Lack of Correlation between Heterozygosity (H) within Species and Genetic Distance (D) between Species

A positive correlation between H and D is taken as an evidence for the neutral theory of evolution as under this theory both H and D are functions of mutation rates. Such a claim based on a pooled set of data from a variety of organisms has indeed been made.[53,54] However a little reflection would suggest that under almost any theory of variation we should expect a positive correlation between H and D but, of course, the strength of correlation will vary between different theories (see Gillespie, in this volume). However, this is not what we see in the *melanogaster* complex species. The D and H show no correlation between *D. melanogaster* and *D. simulans*[7]; in fact a negative trend is suggested by the data as highly heterozygous (multiallelic) loci within species tend to have lower genetic divergence between species. The same relationship holds in a larger data set which includes *D. mauritiana* and *D. sechellia* (Singh and Choudhary, unpublished results). This type of negative trend is expected if most of the genetic divergence between closely related species is due to the partitioning of the genetic polymorphisms from the ancestral species. This is in agreement with what Lewontin[31] said, that "*most of the divergence in the early stages of phyletic evolution makes use of the already available repertoire of genetic variation and is not limited by the rate of appearance of novelties by new mutation.*"

Genetic Divergence as a Function of Tissue Specificity (Function) and Stage of Development

The application of two-dimensional gel electrophoresis (2DE) to protein variation from a variety of reproductive and non-reproductive tissues from *Drosophila* larvae and adults has shown that the level of species divergence is highly tissue specific, i.e., proteins from some tissues, such as brain, show lower levels of divergence and those from the male reproductive tract, higher levels of divergence.[10,57] Lower level of divergence in the early stages of development has also been shown by DNA/DNA hybridization studies in *Drosophila*.[46] The application of native one-dimensional gel electrophoresis has been mostly limited to enzymatic proteins involved in intermediately metabolism and we have had very little knowledge of the amount of genetic variation in other types of proteins involved in early development or in tissue specific functions. If it is true that closely related species tend to differ more in morphological or life history traits which appear in the later stages of life, then we should expect a positive correlation between developmental stage and the levels of genetic divergence. We do not expect closely related species to differ widely in the early stages of their development—genetically or phenotypically.

Genic Divergence of Autosomal Versus Sex-linked Genes

Due to the haploid nature of sex chromosomes, recessive favorable mutations accumulate more rapidly on sex chromosomes than on autosomes.[4] Thus a faster rate of evolution of sex-linked genes than autosomal genes has been proposed to explain the large effect of X chromosome on post-zygotic reproductive isolation and Haldane's rule.[4,18] A higher rate of divergence of sex-linked genes than autosomal genes (if it is true) can also be indicative of the role of sex-related functions in speciation. To test this we applied 2DE analysis to testis proteins of *D. simulans, D. mauritiana* and *D. sechellia* and found that the levels of genetic divergence for X chromosome associated proteins were not higher than those for autosomal proteins (Zeng and Singh[64] and unpublished results). This result is also supported from the large body of gene-enzyme data[6] as well as the DNA sequence variation data.[3] However, a negative result based on the overall levels of divergence in a large number of proteins does not mean that such a difference in rates of divergence might not occur for some specific genes.

Thus to summarize this section, the level and the pattern of genetic divergence between closely related species are highly tissue (function) and developmental stage specific and appear to be molded by population history and natural selection. The level or the pattern of genetic divergence between species do not give any clue to the genetic basis of reproductive isolation, except that the number of genes involved are small.

The Genetic Basis of Post-Zygotic Reproductive Isolation

While the genetic basis of speciation is usually meant to include both the genetic changes that lead to reproductive isolation and changes that lead to species-specific adaptation, the genetic analysis of reproductive isolation between related species is the favored experimental approach to pursue the problem of speciation. Post-zygotic isolation involves well defined characters, i.e., hybrid sterility and inviability, which are especially suitable for genetic investigation. The major results of genetic studies of hybrid sterility and inviability are summarized as follows.

Haldane's Rule and the Large Effects of the X Chromosome

More than 70 years ago, Haldane[24] drew attention to the peculiar nature of species hybrid sterility and inviability and pointed out that "*when in the F_1 offspring of two different animal races one sex is absent, rare, or sterile, that sex is the heterozygous sex.*" This observation is known as Haldane's rule. Systematic classical genetic analysis of hybrid sterility and inviability on various groups of *Drosophila* species pioneered by Dobzhansky[19] and followed by others (reviewed by Coyne and Orr[18]) have revealed a related general pattern in post-zygotic reproductive isolation: *The genes having the greatest effect on hybrid sterility and inviability are X-linked.* These two generalizations are remarkably consistent and are therefore called two rules of speciation.[18] In spite of the consistency of the two rules of speciation, their genetic basis is still unclear.

Since the first formulation of the rule by Haldane,[24] a number of hypotheses have been proposed to explain it (see Wu and Davis[61] for review). Intensive debates on different explanations are still going on[5,45] as none of the proposed hypothesis can completely explain Haldane's rule.[13] It remains a question whether Haldane's rule is a composite rule and there is no unitary explanation as proposed by Wu and Davis,[61] or we simply have not found the right explanation yet.

The empirical generalization of the large X-effect is based on numerous genetic studies of hybrid sterility and inviability in different groups of *Drosophila* including the *obscura*,[19,39,60] the *melanogaster*,[11,17] the *virilis*,[41] and the *repleta* groups.[21,65] These studies, most of which are on hybrid male sterility, show that a replacement of the X chromosome or marked segments of the X chromosome by that of a foreign species usually have a disproportionately greater effect on hybrid male sterility than an autosomal replacement of similar size. However, it is possible that the large X-effect is an artifact due to the nature of Dobzhansky's[19] backcross analysis which has been used in most of these studies.[61] In the backcross analysis, the interspecific replacements of chromosomes or segments of chromosomes are performed on one copy only. The replacements of the X are hemizygous, but the replacements of the autosomes are heterozygous. Both dominant and recessive effects of X chromosome can be revealed by this analysis, but only dominant effects of autosomes can be revealed. In other words the effects of autosomes have not been fully revealed by such an analysis. A more proper analysis would be to compare hemizygous replacements of the X chromosome with homozygous replacements of the autosomes. This type of comparisons have actually revealed large effects of the autosomes as well as of the sex chromosomes in the studies of Hennig[21] and Pantazidis and Zouros.[42] Therefore, the rule of large X-effect needs to be further confirmed by more studies before we try to find any genetic or evolutionary explanation for it.

The Number of Genes Involved in Hybrid Male Sterility

The number of genes involved in hybrid sterility and inviability is one of the most frequently asked questions in the studies of post-zygotic reproductive isolation. The backcross analyses of Coyne,[11] and Coyne and Kreitman[17] have shown that all the five markers they used (each on a major chromosomal arm) is associated with male sterility in species pairs *D. simulans* / *D. mauritiana* and *D. simulans* / *D. sechellia*. As the X chromosome appears to have the highest concentration of sterility genes,[18] Coyne and Charlesworth[15] introgressed three regions of the *D. mauritiana* (or *D. sechellia*) X chromosome into *D. simulans* and mapped three sterility genes using a maximum likelihood method. The authors stressed that this is the maximum number of genes that could be identified and the actual number of genes affecting sterility is probably much greater.

Two of the three regions of *D. sechellia* X chromosome have been further studied by Johnson[28] using molecular marker-assisted introgression method. No additional sterility genes were found. Furthermore, a region comprising approximately 1/4 of the X chromosome of *D. sechellia* (from 13F to 18CD on the polytene chromosome) was found not to possess any hybrid sterility factor, when introgressed into the background of *D. simulans*.[28] These results in combination with our 2D protein data[64] strongly suggest that the number of genes involved in hybrid male sterility in this species pair is not large; instead, a small number of genes may be involved.

The question concerning the number of genes involved in hybrid sterility is only meaningful in relation to the age of the species pair being considered. There may be more genes causing hybrid sterility in older species pairs than in younger species pairs. We must differentiate between the number of genes *initially* required for complete hybrid sterility and the total number of hybrid sterility genes that can be detected in the present day species. Hybrid sterility in young species pairs may involve a very small number of genes and may have a fairly simple genetic basis. This is evident from the studies in *Drosophila pseudoobscura pseudoobscura* and *D. p. bogotana* (Orr[40] and personal communication), in which most chromosomal regions carry no sterility factor and the hybrid male sterility is almost solely due to a small region of the X chromosome.

The simple genetic basis for hybrid inviability has been demonstrated by showing that mutations in single genes can rescue inviable hybrids from crosses between *D. melanogaster* and *D. simulans* (Sawamura and Yamamoto[49] and references therein). Although, it remains to be shown if the genetic basis of hybrid inviability in most closely related species pairs is that simple.

The Nature of Genes Involved in Hybrid Male Sterility

A central question concerning the nature of genes involved in reproductive isolation remains unanswered: *i.e.*, is hybrid sterility or inviability caused by cumulative effect of many minor genes, or by the action of a few major genes, each with a discrete effect? As stated above, the involvement of major genes in hybrid inviability is shown by the isolation of mutations that rescue inviable hybrids.[26,47,48,58] However, for hybrid sterility, the issue remains controversial. Although by using marker-assisted introgression of chromosome segments, a number of major genes have been identified and mapped (see Perez et al. 1993 for references), the introgressed segments are usually large and it is difficult to determine if the effects of the introgressed segments are due to mainly major genes or only to a large number of linked minor genes.

The introgression studies with *D. buzzatii* group species[35,36,37] show that the effects of most introgressed segments are due to large number of genes each with small effect (polygenes). On the other hand, in species pairs *D. simulans* / *D. mauritiana*, and *D. simulans* / *D. sechellia*, Coyne and Charlesworth[16] introgressed three marked regions of *D. mauritiana* (or *D. sechellia*) into *D. simulans* and mapped three major genes on the X chromosome. However, contradictory results have been obtained in the same pair of species for the same introgressed regions. Naveira[38] introgressed two of the three regions studied by Coyne and Charlesworth.[16] Two regions linked to *forked* locus and *yellow-white* loci did not cause sterility when independently introgressed from *D. mauritiana* (or *D. sechellia*) to *D. simulans*, but when introgressed together they caused partial or complete sterility. Naveira[38] concluded that the sterility effects associated with these regions are polygenic, and, therefore raised the possibility that most of the previously identified "major genes"[15,40,42] may be polygenes. Similar introgressions in these three species have been also carried out in Wu's lab (Chung-I Wu, personal communication) and similar results have been obtained.

It is possible that the regions around the *forked* locus and the *yellow-white* loci carry both major genes and polygenes and different studies introgressed different subregions which carry different types of genes. Although it is difficult to definitely prove the existence of a major gene, all "major" genes can not be blocks of polygenes. Why should all the polygenes causing sterility be clustered in small regions? Polygenic mutations causing incompatibility between sibling species are expected to occur randomly in the genome. The probability of co-occurrence of a number of incompatible mutations causing complete sterility in a very small region of a chromosome would be very small.

The data published so far suggest that both major and minor genes are involved in hybrid male sterility. A general method for detecting major hybrid male sterility genes, described in Zeng and Singh,[65] can be used in any species pair which produce unisexual hybrid sterility (which is the most common form of reproductive isolation between closely related animal species). One way to prove that an identified "major" gene is a major gene is to screen for mutants which can restore fertility of males carrying the sterility gene. Such studies would also provide information on the nature of the action of these genes.

Asymmetric Epistatic Interactions Underlying Hybrid Sterility and Inviability

Hybrid sterility or inviability results when two sets of genes from different species are brought together into hybrid individuals. As each of the two sets of genes function normally in the parental species the incompatibility (hybrid sterility or inviability) must necessarily be a result of epistatic interactions of certain genes from the two species. There are many possible incompatible interactions between heterospecific chromosomes, and most of them have been observed. The most common form of interaction is the X-autosome interaction which has been observed in many pairs of *Drosophila* species (Zeng and Singh[63] and the references cited in there). In pairs of *D. hydei* / *D. neohydei,*[50] interactions between different autosomes and between the Y chromosome and autosomes have also been found. Interactions between Y chromosome and autosomes have also been observed in species pair *D. virilis* / *D. lummei.*[25] A well established case of interactions between autosomes and Y chromosome is in the species pair *D. mojavensis* and *D. arizonensis*. Interaction between heterospecific sex chromosomes has also been reported.[12,39]

The incompatible interactions between genes from different species are often *asymmetrical* and this pattern has been revealed in most species pairs where data regarding sterility interactions are available.[60,62,65] For example, the *se-sh* region of *D. pseudoobscura* X chromosome is compatible with the genetic background of *D. persimilis* whereas the reciprocal introgression results in male sterility.[60] The asymmetrical interactions (asymmetrical X-autosome interaction or Y-autosome interaction) explain the often observed *unidirectional* hybrid sterility and inviability (nonreciprocal hybrid sterility and inviability). By extrapolation, the asymmetrical model can also account for the commonly observed unidirectional mating (mating between two species in one direction is easier than in the reciprocal direction) in a similar way. The asymmetrical nature of the interactions and the genetic and population processes that lead to it may provide a key to the understanding of the evolution of reproductive isolation between animal species.

The Genetic Basis of Speciation

Causative and Non-Causative Theories of Speciation

We must differentiate between causative and non-causative theories of speciation, i.e., between theories that can tell us about the nature of genetic changes that lead to speciation and those that can not. So, for example, Goldschmidt's[23] macromutation theory of speciation is not a causative theory as the macromutations are simply involved to explain large gaps between species, and similarly the role of chromosomal changes in speciation is involved to explain the observed fact that most species differ in their karyotypes. In this respect, the large effect of X-chromosome on hybrid sterility and inviability is more of a causative rule than Haldane's rule which is simply a general statement about reproductive isolation—an *outcome* of speciation.

The causative theories of speciation should be predictive about the nature of divergence during the speciation process and they should be able to explain the differential nature of genetic divergence in *sexual* genes and traits that are characteristically involved in the development of reproductive isolation. There are three general observations that we think together provide a basis for a causative theory of speciation. These are (1) that sex chromosomes (X or Y) appear to be always involved in affecting male sterility,[13] (2) that genetic interaction underlying sterility and inviability in species hybrids are often asymmetric (unidirectional sterility),[63] and (3) that sexual traits in general, ranging from molecules to morphology, appear to show more divergence than non-sexual traits.[57] Based on these observations the two most general characteristics of a causative theory of speciation are *asymmetric genetic interactions* and *differential divergence of sexual and non-sexual traits.*[51,52]

Towards a Unified Theory of Speciation

All currently available models of speciation (Table 2) feature some combination of genetic, populational, and ecological factors, and the models differ depending upon which factors they emphasize the most. So the geographic models of speciation emphasize isolation, the founder effect models population bottlenecks, and the genomic models, major changes in the genome arising from within (e.g., developmental macromutations) or from without (e.g., transposable elements). Since each of these model was developed with a particular group of organisms

Table 2. Genetic models of speciation

1. Sympatric model: Adaptive divergence
2. Allopatric model: Geographic isolation, founder effect and selection
3. Founder-flush model: Divergence of polygenic complex, divergence of mating behavior
4. Genetic transilience model: Divergence of major genes and their modifiers, divergence of mating behavior
5. Genomic disease model: Loss or gain of transposable elements
6. Mechanical genome incompatibility model: Chromosomal divergence, problems with chromosome pairing
7. Macromutation model

Table 3. A two-component gene pool model of adaptation and speciation

Genetic/gene pool	Survival/adaptation	Reproduction/speciation
1a. Brain/behavior	Behavioral adaptation	Premating isolation
1b. Gametogenesis	—	Post-zygotic reproductive isolation
2. Development and physiology	Morphological/physiological adaptation	Reproductive isolation as a result of pleiotropic effect

in mind, each can be said to contain some element of truth but certainly not the whole truth. The geographic models of speciation were developed with highly mobile organisms in mind, the founder effect models with role of behavioral trait in mind, and the genomic models with role of macromutations (affecting many traits at once) and rapid rates of speciation. Therefore it is not surprising that none of the proposed theories of speciation satisfactorily apply to all organisms or even to most organisms. No single theory of speciation can be both general enough to apply to all organisms and specific enough to provide a detailed picture of speciation in a given group of organisms. A general genetic theory of speciation must be able to do two things. It must make statements about the differential nature of genetic divergence of sexual and non-sexual trait, and it must contain all forces of evolutionary change, preferably with particular reference to the groups of organisms to which these forces are individually more likely to apply.

The first criterion can be fulfilled by reviving Dobzhansky's[20] "gene pool" concept of speciation in which species are defined on the basis of their actual or potential "genome incompatibility" and not on the basis of sexual reproduction, morphological identification, geographical lineage or ease of practical application. If the "gene pool incompatibility" can be taken as a central criterion of speciation, then we can think of the species gene pool as being made of two loosely coupled sub-gene pool systems—one involved in organisms' *survival* and the other in *reproduction* (Table 3). The gene pool system involved in survival can be defined as comprised of all genes that are essential for normal development and differentiation leading to mature adults. The gene pool system involved in reproduction can be defined as comprising of those genes that are involved in the development of germline, genitalia and mating behavior, all of which are required for a successful sexual reproduction. While there may be large number of genes that affect both types of trait, i.e., survival and reproduction, there must be a group of genes whose primary role is in reproduction. The latter group of genes must affect reproductive isolation directly by way of affecting *mating behavior* or *gametogenesis,* while the former group of genes affecting survival must affect reproductive isolation only indirectly as a byproduct of adaptive or non-adaptive genetic divergence.

With respect to the role of evolutionary forces causing genetic divergence no theory of speciation can be satisfactory without giving some emphasis to isolation in animals, to adaptive divergence in plants, and to mutational divergence in bacteria. During the last thirty years no progress has been made in the field of speciation simply by focusing on the populational aspects of speciation. The current emphasis on the genetic analysis of premating[14] and post-zygotic reproductive isolation is a sign of rapid progress to come in the near future. Complete details of the unified theory of speciation will be published elsewhere (Singh, in preparation).

References

1. F. J. Ayala. (1975). Genetic differentiation during the speciation process. *Evol. Biol.* **8,** 1–78.
2. W. Bateson. (1894). *Materials for the study of variation*. Macmillan, London.
3. D. J. Begun and C. F. Aquadro. (1992). Levels of naturally occurring DNA polymorphism correlate with recombination rates in *D. melanogaster*. *Nature* **356,** 519–520.
4. B. Charlesworth, J. A. Coyne and N. H. Barton. (1987). Relative rates of evolution of sex chromosomes and autosomes. *American Naturalist* **130,** 113–146.
5. B. Charlesworth, J. A. Coyne and H. A. Orr. (1993). Meiotic drive and unisexual hybrid sterility: a comment. *Genetics* **133,** 421–424.
6. M. Choudhary and R. S. Singh. (1987). A comprehensive study of genic variation in natural populations of *Drosophila melanogaster* III. Variation in genetic structure and their causes between *Drosophila melanogaster* and its sibling species *Drosophila simulans*. *Genetics* **117,** 697–710.
7. M. Choudhary, M. B. Coulthart and R. S. Singh. (1992). A comprehensive study of genic variation in natural populations of *Drosophila melanogaster* VI. Patterns and processes of genic divergence between *D. melanogaster* and its sibling species *D. simulans*. *Genetics* **130,** 843–853.
8. M. B. Coulthart and R. S. Singh. (1988a). Differing amounts of genetic polymorphism in testes and male accessory glands of *Drosophila melanogaster* and *Drosophila simulans*. *Biochem. Genet.* **26,** 153–153.
9. M. B. Coulthart and R. S. Singh. (1988b). High level of divergence of male-reproductive-tract proteins, between *Drosophila melanogaster* and its sibling species, *D. simulans*. *Mol. Biol. Evol.* **5,** 182–182.

10. M. B. Coulthart and R. S. Singh. (1988c). Low genic variation in male-reproductive-tract proteins of *Drosophila melanogaster* and *D. simulans*. *Mol. Biol. Evol.* **5,** 167–167.

11. J. A. Coyne. (1984). Genetic basis of male sterility in hybrid between two closely related species of *Drosophila*. *Proc. Natl. Acad. Sci. U.S.A.* **81,** 4444–4447.

12. J. A. Coyne. (1985). The genetic basis of Haldane's rule. *Nature* **314,** 736–738.

13. J. A. Coyne. (1992). Genetics and speciation. *Nature* **355,** 511–515.

14. J. A. Coyne. (1993). The genetics of an isolating mechanism between two sibling species of *Drosophila*. *Evolution* **47,** 778–788.

15. J. A. Coyne and B. Charlesworth. (1986). Localization of an X-linked factor causing sterility in male hybrids between two closely related species of *Drosophila*. *Heredity* **57,** 243–246.

16. J. A. Coyne and B. Charlesworth. (1989). Genetic analysis of X-linked sterility in hybrids between three sibling species of *Drosophila*. *Heredity* **62,** 97–106.

17. J. A. Coyne and M. Kreitman. (1986). Evolutionary genetics of two sibling species, Drosophila simulans and *D. sechellia*. *Evolution* **40(4),** 673–691.

18. J. A. Coyne and H. A. Orr. (1989). Two rules of speciation. pp. 189–211 in *Speciation and its consequences,* edited by D. Otte and J. Endler. Sinauer Press, Sunderland, Mass.

19. Th. Dobzhansky. (1936). Studies on hybrid sterility. II. Localization of sterility factors in *Drosophila pseudoobscura* hybrids. *Genetics* **21,** 113–135.

20. Th. Dobzhansky. (1937a). *Genetics and The Origin of Species,* Columbia University Press, New York.

21. W. Hennig. (1977). Gene interactions in germ cell differentiation of *Drosophila*. Pages 363–371 in G. Weber, ed. *Advances in enzyme regulations*. Vol. 15. Pergamon, Oxford.

22. J. Gillespie. (1991). *The Causes of Molecular Evolution*. Oxford, London.

23. R. Goldschidt. (1940). *The material basis of evolution*. Yale University, New Haven.

24. J. B. S. Haldane. (1922). Sex ratio and unisexual sterility in hybrid animals. J. *Genetics* **12,** 101–109.

25. E. Heikkinen and J. Lumme. (1991). Sterility of males and female hybrids of *Drosophila virilis* and *Drosophila lummei*. *Heredity* **67,** 1–11.

26. P. Hutter and M. Ashburner. (1987). Genetic rescue of inviable hybrids between Drosophila melanogaster and its sibling species. *Nature* **327,** 331–333.

27. J. S. Huxley. (1942). *Evolution: The modern Synthesis*. London: Allen & Unwin.

28. N. A. Johnson. (1992). Genetics of hybrid male sterility in three sibling species of the *Drosophila melanogaster* species subgroup. Ph.D. thesis, University of Rochester, Rochester, New York.

29. N. A. Johnson and C.-I. Wu. (1992). An empirical test of the meiotic drive models of hybrid sterility: sex-ratio data from hybrids between *Drosophila simulans* and *Drosophila sechellia*. *Genetics* **130,** 507–511.

30. R. M. Kliman and J. Hey. (1993). DNA sequence variation at the period locus within and among species of the *Drosophila melanogaster* complex. *Genetics* **133,** 375–387.

31. R. C. Lewontin. (1974). *The genetic basis of evolutionary change*. Columbia, New York.

32. E. Mayr. (1963). *Animal Species and Evolution*. Harvard, Cambridge.

33. E. Mayr. (1982). *The Growth of Biological Thought: Diversity, Evolution and Inheritance*. Harvard, Cambridge.

34. M. Nei. (1987). *Molecular Evolutionary Genetics*. Columbia, New York.

35. H. Naveira and A. Fontdevila. (1986). The evolutionary history of *Drosophila buzzatii*. XII. The genetic basis of sterility in hybrids between *D. buzzatii* and its sibling *D. serido* from Argentina. *Genetics* **114,** 841–857.

36. H. Naveira and A. Fontdevila. (1991a). The evolutionary history of *Drosophila buzzatii*. XXI. Cumulative action of multiple sterility factors on spermatogenesis in hybrids of *D. buzzatii* and *D. koepferae*. *Heredity* **67,** 57–72.

37. H. Naveira and A. Fontdevila. (1991b). The evolutionary history of *Drosophila buzzatii*. XXII. Chromosomal and genic sterility in male hybrids of *D. buzzatii* and *D. koepferae*. *Heredity* **66,** 233–240.

38. H. F. Naveira. (1992). Location of X-linked polygenic effects causing sterility in male hybrids of *Drosophila simulans* and *D. mauritiana*. *Heredity* **68,** 211–217.

39. H. A. Orr. (1987). Genetics of male and female sterility in hybrids of *Drosophila pseudoobscura* and *D. persimilis*. *Genetics* **116,** 555–563.

40. H. A. Orr. (1989). Localization of genes causing postzygotic isolation in two hybridizations involving *Drosophila pseudoobscura*. *Heredity* **63,** 231–237.

41. H. A. Orr and J. A. Coyne. (1989). The genetics of postzygotic isolation in the *Drosophila virilis* group. *Genetics* **121,** 527–537.

42. A. C. Pantazidis and E. Zouros. (1988). Location of an autosomal factor causing sterility in *Drosophila mojavensis* males carrying the *Drosophila arizonensis* Y chromosome. *Heredity* **60,** 299–304.

43. A. C. Pantazidis, V. K. Galanopoulos and E. Zouros. (1993). An autosomal factor from *Drosophila arizonae* restores normal sptermatogenesis in *Drosophila mojavensis* males carrying the *Drosophila arizonae* Y chromosome. *Genetics* **134,** 309–318.

44. D. E. Perez, C.-I. Wu, N. A. Johnson and M.-L. Wu. (1993). Genetics of reproductive isolation in the Drosophila clade: DNA marker-assisted mapping and characterization of a hybrid-male sterility gene, *Odysseus (Ods)*. *Genetics* **133,** 261–275.

45. A. Pomiankowski and L. D. Hurst. (1993). Genomic conflicts underlying Haldane's rule. *Genetics* **133,** 425–432.

46. J. R. Powell, A. Caccone, J. M. Gleason and L. Nigro. (1993). Rates of DNA evolution in Drosophila depend on function and developmental stages of expression. *Genetics* **133,** 291–298.

47. K. Sawamura, T. Taira and T. K. Watanabe. (1993). Maternal hybrid rescue (mhr):

the gene which rescues embryonic lethal hybrids from *Drosophila simulans* females crossed with *D. melanogaster* males. *Genetics* **133,** 299–305.

48. K. Sawamura, M.-T. Yamamoto and T. K. Watanabe. (1993). Hybrid lethal systems in Drosophila melanogaster species complex. II. The *zygotic hybrid rescue (Zhr)* gene of *D. melanogaster*. *Genetics* **133,** 307–313.
49. K. Sawamura and M.-T. Yamamoto. (1993). Cytogenetical localization of Zygotic hybrid rescue (Zhr), a Drosophila melanogaster gene that rescues interspecific hybrids from embryonic lethality. *Mol Gen Genet* **239,** 441–449.
50. U. Schafer. (1978). Sterility in *Drosophila hydei* X *Drosophila neohydei* hybrids. *Genetica* **49,** 205–214.
51. R. S. Singh. (1989). Population genetics and evolution of species related to *Drosophila melanogaster*. *Ann. Rev. Genet.* **23,** 425–453.
52. R. S. Singh. (1990). Patterns of species divergence and genetic theories of speciation. In K. Wohrman and S. K. Jain (eds.), *Population Biology: Ecological and evolutionary viewpoints,* pp. 231–264, Springer-Verlag, Berlin, Hadelberg.
53. D. O. F. Skibinski and R. D. Ward. (1981). Relationship between allozyme heterozygosity and rate of divergence. *Genet. Res.* **38,** 71–92.
54. D. O. F. Skibinski and R. D. Ward. (1982). Correlations between heterozygosity and evolutionary rates of proteins. *Nature* **298,** 490–492.
55. G. L. Stebbins. (1950). *Variation and Evolution*. Harvard, Cambridge.
56. L. H. Throckmorton. (1977). *Drosophila* systematics and biochemical evolution. *Ann. Rev. Ecol. Syst.* **8,** 235–254.
57. S. Thomas and R. S. Singh. (1992). A comprehensive study of genic variation in natural populations of *Drosophila melanogaster* VII. Varying rates of genic divergence as revealed by 2-dimensional electrophoresis. *Mol. Biol. Evol.* **9,** 507–525.
58. T. K. Watanabe. (1979). A gene that rescues the lethal hybrids between *Drosophila melanogaster* and *D. simulans*. *Japanese Journal of Genetics* **54,** 325–331.
59. M. J. D. White. (1978). *Modes of Speciation*. Freeman, San Francisco.
60. C.-I. Wu and A. Beckenbach. (1983). Evidence for extensive genetic differentiation between the sex ratio and the standard arrangement of *Drosophila pseudoobscura* and *D. persimilis* and identification of hybrid sterility factors. *Genetics* **105,** 71–86.
61. C.-I. Wu, and A. W. Davis. (1993). Evolution of postmating reproductive isolation: The composite nature of Haldane's rule and its genetic bases. *American Naturalist* **142,** 187–212.
62. C.-I. Wu, D. E. Perez, A. W. Davis, N. A. Johnson, E. L. Cabot. M. F. Palopoll and M.-L. Wu. (1992). Molecular genetic studies of postmating reproductive isolation in *Drosophila*. In: *Proceedings of the 17th Taniguchi Symposium in Biophysics,* Mishima, Japan, edited by N. Takahata and A. G. Clark. Springer-Verlag, Berlin.
63. L.-W. Zeng and R. S. Singh. (1993a). The genetic basis of Haldane's rule and the nature of asymmetric hybrid male sterility between *Drosophila simulans, D. mauritana* and *D. sechellia*. *Genetics* **134,** 251–260.
64. L.-W. Zeng and R. S. Singh. (1993b). A combined classical genetic and high

resolution two-dimensional electrophoretic approach to the assessment of the number of genes affecting hybrid male sterility in *Drosophila simulans* and *Drosophila sechellia*. *Genetics* **135,** 135–147.

65. L.-W. Zeng and R. S. Singh. (1994). A general method for the identification and mapping of major genes affecting hybrid male sterility in *Drosophila*. *Proc. Natl. Acad. Sci., USA* (submitted).

66. E. Zouros, K. Lofdahl, and P. Martin. (1988). Male hybrid sterility in *Drosophila:* interactions between autosomes and sex chromosomes of *D. mojavensis* and *D. arizonensis*. *Evolution* **42,** 1321–1331.

19

Polymorphism at *Mhc* Loci and Isolation by the Immune System in Vertebrates

Naoyuki Takahata

Abstract. *The major histocompatibility complex (Mhc)* contains a family of genes whose products play crucial roles in self and nonself discrimination. Self is the mature T cell repertoire that is restricted by Mhc molecules, whereas nonself is processed peptides of foreign antigens that can be presented in the context of Mhc. Because the T cell repertoire and the complex of Mhc-peptides provide a molecular basis of self defense mechanism and both of them depend critically on *Mhc* genotypes, the genes must be subjected to natural selection to some extent. A model of individual fitness is formulated in terms of the number of functional *Mhc* loci, the number of heterozygous loci, the size of T cell repertoire, and the size of Mhc peptide repertoire. Available data delimit the range of values of these parameters in the model. The mode of natural selection on *Mhc* loci depends primarily on the T cell repertoire shaped by different gene products. If each *Mhc* gene product deletes a substantial portion of T cells, balancing selection cannot operate simultaneously on a large number of loci. Only when the number of functional *Mhc* loci is limited, natural selection acts so as to produce the unusual polymorphism. However, if individuals are heterozygous at all these loci or possess haplotypes being different in the constellation of *Mhc* loci, their fitness may be reduced. The reduction may happen in hybrids between antigenically diverged populations, in which case the immune system acts as a reproductive isolation mechanism.

Introduction

Molecules encoded by the major histocompatibility complex (*Mhc*) loci shape the T cell repertoire in the thymus (Mhc-restriction) and present processed peptides of foreign antigens to T cell receptors (Tcr), thereby playing a critical role in the cellular immune response.[1] Since different Mhc molecules have different specificities in T cell recognition and peptide binding,[2] it would be surprising if this dual function of Mhc molecules has nothing to do with natural selection. In fact, recent molecular studies demonstrate the operation of natural selection, although the precise evolutionary mechanism is still a matter of debate.[3–5]

Like *Tcr* and immunoglobulin (*Ig*) genes, *Mhc* genes proper form a multigene family.[6,7] The human genome contains about 30 *HLA* loci, 10 class I α loci and some 20 class II α and β loci,[8] although not all are functional.[7] In the mole rat genome, only two or three out of more than 60 class I *Mhc* loci are expressed to the extent that their products can carry out the function of antigen presentation.[9] The limit of the number of functional *Mhc* loci is usually accounted for in relation to the T cell repertoire restricted by Mhc.[7,10,11] It is intuitively comprehensible that too small a T cell repertoire would weaken the defense.

Unlike *Tcr* and *Ig* genes, *Mhc* genes do not undergo somatic rearrangement of the genome to generate their diversity.[6,7] It is argued that Mhc molecules must be expressed *nonclonally* in order to ensure their sufficient density on all the antigen-presenting cells and for this need, the *Mhc* diversity cannot be generated by somatic rearrangement; it is only useful when clonal expansions of cells are allowed.[10] Instead, the diversity of Mhc molecules is manifest when individuals in a population are compared. Some of functional *Mhc* loci are strikingly polymorphic in populations,[7] provided that they have not undergone severe bottlenecks, as exemplified by the cheetah.[12] To understand these features of Mhc molecules and the vertebrate immune system ultimately, it is necessary to understand quantitatively the thymic selection and natural selection concerning Mhc molecules.

There is another interesting possibility that Mhc molecules might play a role in the evolution of vertebrates, which was first suggested by Haldane[13] (see also ref. 14). Haldane's papers abound in speculations, and naturally many of them have turned out to be wrong, but much more remarkable are the ones that have turned out to be on target and path-breaking.[15,16] Among his many speculations, it seems to me that the most fascinating is this: *Once a pair of races is geographically separated they will be exposed to different pathogens. Such races will tend to diverge antigenetically, and some of this divergence may lower the fertility of crosses.*[13] Since major self-antigens are *Mhc*-coded, what was suggested is that divergence of *Mhc* loci, either through polymorphism or gene duplication, lowers the fertility of hybrids. There is no experiment to prove or disprove this theme of speciation. Speciation may involve changes of many molecules and depend on the way they communicate. Elucidation of these molecules or their communication is certainly a challenging task to evolutionary biologists. However, since such information is virtually absent, I take a reverse approach. I ask whether or not the well-studied Mhc moleclues or the tri-molecular communication among Mhc, Tcr and peptides has something to do with speciation. Based on a fitness model conferred by the tri-molecular communication, I explain some of the evolutionary features of Mhc and examine the tri-molecular communication as being important in speciation.

A Single Locus Model

Since the process of thymic selection is not fully elucidated,[17,18] it is necessary to make some assumptions about positive and negative selection as well as about the involvement of self peptides in positive selection. Positive selection may take place at an early stage of T-cell development and maximize the size of T-cell clones,[18] but the early occurrence of negative selection is also observed.[19] When immunological tolerance is induced by eliminating or suppressing autoreactive T-cell clones, self peptides must be involved, but this is not clarified as regards positive selection.[2] It is assumed that positive selection occurs before negative selection and that self peptides are necessitated in positive selection. These assumptions seem reasonable, because if negative selection does not follow positive selection, there may remain autoreactive T cells and because an Mhc molecule is unstable without binding any peptide.[20]

Before introducing a model of multiple loci, we consider a fitness model of single locus to define necessary quantities. *Mhc* alleles are expressed codominantly in individuals [6]. In positive selection, different *Mhc* allelic products may mould different sets of naive T-cell clones. The extent of overlap in T-cell clones depend on differences of Mhc molecules in their amino acid sequences. Thus, if homozygotes mould a certain set of naive T-cell clones, heterozygotes mould two such sets which may or may not overlap from each other. For generality, denote by a ($0 \leq a \leq 1$) the extent of nonoverlap for a randomly chosen pair of *Mhc* alleles. If the size of naive T-cell clones in homozygotes is r, it is $r(1 + a)$ in heterozygotes. Heterozygotes possibly possess twice as many naive T-cell clones as homozygotes ($a = 1$). When self-peptides are involved in positive selection and if there is a synergistic interaction in forming the Mhc-peptide complex, a is likely to be strictly positive. The r and $r(1 + a)$ are assumed to be applied to all kinds of homozygotes and heterozygotes when there are more than two alleles segregating in a population. If there are substantial differences in the size of T-cell repertoires among homozygotes or among heterozygotes, the resulting evolutionary force would be directional and may lead to monomorphism in an extreme situation. This possibility is excluded in the following.

To induce immunological tolerance, some naive T-cell clones that react to the complex of Mhc and self peptides must be deleted. This results in the so-called blind spot in the mature T-cell repertoire.[7] Let b be the proportion of blind spot produced by one type of Mhc molecules ($0 \leq b \leq 1$). The size of the mature T-cell repertoire of homozygotes is then given by $r(1 - b)$. In heterozygotes, the other type of Mhc molecules may further delete some T-cell clones. Denote by d ($0 \leq d \leq b$) such an excess of blind spot in heterozygotes. The size of the mature T-cell repertoire of heterozygotes is given by $r(1 + a)(1 - b)(1 - d)$.

Each *Mhc* allelic product has a certain specificity for binding nonself peptides. Let p be the size of peptide repertoire of one type of Mhc molecules. If there are differences in the peptide binding region of different Mhc molecules, they can bind different sets of nonself peptides.[21–23] Hence heterozygotes may have a larger size of peptide repertoire than homozygotes. An excess in the peptide repertoire of heterozygotes is defined as c ($0 \leq c \leq 1$). The peptide repertoire size is given by p for homozygotes and $p(1 + c)$ for heterozygotes.

Fitness of individuals must be somehow related to the size of mature T-cell repertoire as well as the size of peptide repertoire of Mhc molecules. It seems reasonable to assume that the larger these sizes, the higher the survival probability of individuals and that the fitness value is the product of both repertoire sizes; $rp(1 - b)$ for homozygotes and $rp(1 + a)(1 + c)(1 - b)(1 - d)$ for heterozygotes. We can still argue that the difference is irrelevant to natural selection, but for this argument to be valid, we must postulate very special forms of fitness function or particular relationships among the parameters. Since neither is likely, natural selection should operate. It is immediately clear that the selection is a form of either overdominance (heterozygote advantage) or underdominance (heterozygote disadvantage), depending on whether $(1 + a)(1 + c)(1 - d)$ is larger or smaller than 1. Were there no excess of blind spot in heterozygotes ($d = 0$), overdominance would always occur.

A multiple locus model

It is necessary to extend the above model to multiple *Mhc* loci. In the class I, the appropriate number of *Mhc* loci is the number of α loci whereas in the class II, the product of the numbers of α and β loci in each subclass is summed up over all subclasses. Since the peptide repertoires of *HLA-A* and *-B* are largely nonoverlapping,[24] we assume that the same is true for other functional loci (including monomorphic loci). Suppose that there are m such *Mhc* loci and that an individual is heterozygous at n loci ($0 \leq n \leq m$) and homozygous at the remaining $m - n$ loci. The size of peptide repertoire of this individual is given by $np(1 + c) + (m - n)p = p(m + nc)$. In thymic selection, different *Mhc* loci produce nonoverlapping T-cell clones. The size of naive T-cell repertoire is expressed as $nr(1 + a) + (m - n)r = r(m + na)$. When negative selection takes place, m *Mhc* loci delete m sets of T-cell clones independently from the total pool. The remaining proportion is $(1 - b)^m$. In addition, n heterozygous loci further reduce the repertoire size by $(1 - d)^n$. The proportion of mature T-cell repertoire is $(1 - b)^m(1 - d)^n$. The relative fitness of the individual is therefore given by $rp(m + na)(m + nc)(1 - b)^m(1 - d)^n$ (Table 1). We may ignore the factor rp common to all individuals. For convenience, we define $\theta_0 = -\ln(1 - b) \geq 0$ and $\theta_1 = -\ln(1 - d) \geq 0$ ($\theta_0 \geq \theta_1$) and rewrite the fitness function as

$$f(n,m) = C(m + na)(m + nc)\, e^{-(m\theta_0 + n\theta_1)}$$

where C is a scaling constant so that the maximum value of $f(n,m)$ is 1.

First, we note that a large number of loci cannot be polymorphic as long as $d \neq 0$ or $\theta_1 \neq 0$. Because the condition that some of m loci is polymorphic is $a + c > m\theta_1$ (see legend of Table 1), none of m loci can be polymorphic if m is larger than $(a + c)/\theta_1$. Denote this minimum number by m_1, which is given by

Table 1. Thymic and natural selection associated with tolerance induction and peptide presentation of Mhc molecules.

	Thymic Selection		Mhc	Natural Selection
Genotypes	Postive	Negative	Peptide Repertoire	Relative Fitness
Heterozygous at n loci and homozygous at m − n loci	r(m+an)	r(m+na) × $(1-b)^m(1-d)^n$	p(m+nc)	rp(m+na) × (m+nc) × $(1-b)^m(1-d)^n$

m = number of functional *Mhc* loci, r = the size of T-cell clones moulded by one type of Mhc molecules, a = an excess of the size in T-cell clones that are moulded by different types of Mhc molecules coded by the same locus, b = the size of blind spot due to a single kind of Mhc, d = an excess of the blind spot due to a heterozygous *Mhc* locus, p = the size of pathogen epitopes, c = the extent of nonoverlap in peptide presentation between different *Mhc* allelic products. Different *Mhc* loci are assumed to be independent, nonoverlapping in Mhc-restriction and peptide presentation: Peptide presentation and the effect of positive selection are additive while the effect of negative selection is multiplicative. The present model does not distinguish selective advantages in peptide presentation from those in positive selection. Since an Mhc molecule can bind peptides derived from various pathogens, the size of pathogen epitopes is assumed to be constant and independent of *Mhc* genotypes. However, some models for coevolution between Mhc and pathogens (e.g., ref. 33) assume the specific interaction in which case p depends on frequencies of *Mhc* alleles and pathogens.

For a given m, we consider the inequality of $f(n+1,m) > f(n,m)$ for some n, which is written as $-\ln[1 - a/\{m(1 + ax)\}]\,[1 - c/\{m(1 + cx)\}] > \theta_1$ where $x = n/m$. The left hand side decreases monotonously from $-\ln(1 - a/m)(1^{-} c/m)$ to $-\ln[1 - a/\{m(1 + a)\}]\,[1 - c/\{m(1 + c)\}]$. Therefore, if $-\ln[1 - a/\{m(1 + a)\}]\,[1 - c/\{m(1 + c)\}] < \theta_1 < -\ln(1 - a/m)(1 - c/m)$, $f(n,m)$ increases up to some $n < m$. For $-\ln[1 - a/\{m(1 + a)\}][1 - c/\{m(1 + c)\}] > \theta_1$, $f(n,m)$ takes the maximum value at $n = m$. Thus, as long as $-\ln(1 - a/m)(1 - c/m) > \theta_1$, heterozygotes are more advantageous than homozygotes. When a/m and c/m are small, the consition is $a + c > m\theta_1$ or $m_1 > m$.

For a fixed value $x = n/m$ $(0 \le x \le 1)$, the fitness function $f(xm,m)$ is maximized at $m_x = 2/(x\theta_1 + \theta_0) < m_0 = 2/\theta_0$. However, $f(xm_x,m_x) \le f(0,m_0)$ holds if $a + c \le 2\theta_1/\theta_0 = m_0\theta_1$ (or $m_1 < m_0$). The ratio of $f(xm_x,m_x)$ to $f(0,m_0)$ equals to $R(x) = (1 + ax)(1 + cx)/(1 + x\theta_1/\theta_0)^2$. When $a + c \le 2\theta_1/\theta_0$, $ac < (\theta_1/\theta_0)^2$ so that $R(x) < 1$ for any x. Thus, if $a + c \le 2\theta_1/\theta_0$, polymorphism does not evolve in the presence of individuals with m_0 homozygous loci. If on the other hand $a + c > 2\theta_1/\theta_0$, $R(x) > 1$ for some $x > 0$, i.e., there always exist some heterozygotes whose fitness $f(xm_x,m_x)$ is higher than $f(0,m_0)$. Since $f(0,m_x) < f(0,m_0)$, $f(0,m_x) < f(xm_x,m_x)$ is automatically met.

$$m_1 = \frac{a + c}{\theta_1} \tag{1}$$

if $a + c > \theta_1$, and $m_1 = 1$ if $a + c \le \theta_1$. If the number of functional *Mhc* loci in all haplotypes in a population is greater than m_1, all *Mhc* loci should be monomorphic. This situation is related to the so-called permanent heterozygosity[25] in which the demanded genetic diversity is presumed to be supplied by functionally diverged duplicated loci rather than by polymorphism. There is an optimum number of homozygous loci with which the fitness function of homozygotes is maximized. This number is given by

$$m_0 = \frac{2}{\theta_0} \text{ or } (1 - b)^{m_0} = e^{-2} \tag{2}$$

if $2 \geq \theta_0$ ($b \leq 1 - e^{-2} = 0.86$) and $m_0 = 1$ if $2 < \theta_0$. Since $(1 - b)^{m_0}$ is the size of T-cell repertoire, this formula implies that 86% of T-cell clones are to be deleted in the thymus if individuals are monomorphic at m_0 functionally diverged loci.

For the model to be consistent with *Mhc* polymorphism data, m_0 and m_1 must be larger than 1. Whether $m_0 < m_1$ or $m_0 > m_1$, however, depends on three values of $a + c$, θ_0, and θ_1. If $m_0 \geq m_1$ or $a + c \leq 2\theta_1/\theta_0$, fitness of homozygotes at all m_0 *Mhc* loci becomes highest among all possible genotypes (Table 1). In this case, disadvantages due to blind spots are so strong as to prohibit the evolution of polymorphism at any loci. In the opposite case of $m_0 < m_1$, there exist individuals heterozygous at some of m ($\leq m_0$) loci whose fitness becomes higher than that of individuals homozygous at all m_0 loci (Table 1 and Figure 1). The possible existence of hemizygotes in a population does not change these results, because hemizygous loci are equivalent to homozygous loci in the present fitness model of codominance.[6]

Evaluation of Parameters

Available information allows us to evaluate the parameter values in the model. Assume that the number of functional *Mhc* loci in diploid species is m. First, the proportion (g) of T-cell clones purged in the thymus may be more than 90% (ref. 26), because only 2% of T-cells migrate into periphery every day and 98% die in the thymus.[1] Suppose that this observation of g is made for individuals with n heterozygous and $m - n$ homozygous loci. The value of $1 - g$ corresponds to $(1 - b)^m(1 - d)^n = e^{-(m\theta_0 + n\theta_1)}$, although m and n may or may not be known. We have

$$m\theta_0 + n\theta_1 = -\ln(1 - g),$$

the right hand side being 2.3 for $g = 0.90$ and 3.9 for $g = 0.98$. Because there are a number of monomorphic loci ($m >> n$) and $\theta_0 > \theta_1$ by definition, the major contribution to g appears to come from different loci rather than allelic products. It follows that

$$m\theta_0 \approx -\ln(1 - g) \tag{3a}$$

approximately. If $g = 0.9$ and m is known, we can estimate the value of b as

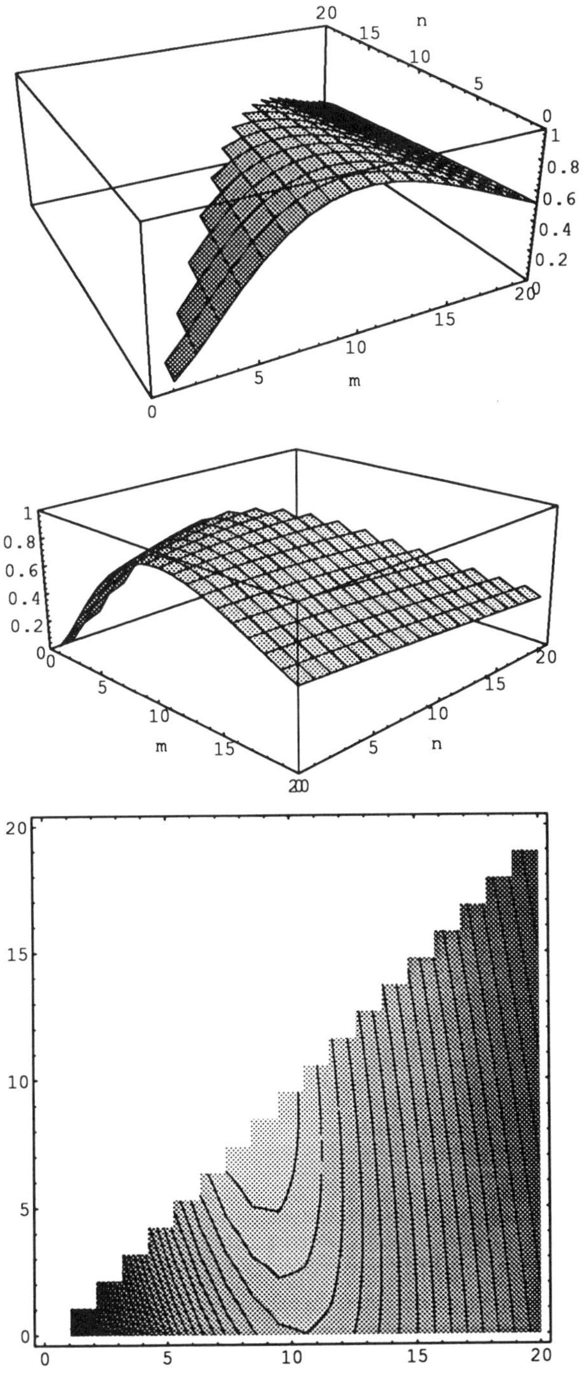

Figure 1. Relative fitness surface (top and middle) for the various numbers of functional *Mhc* loci (m) and heterozygous loci (n). The contour map of fitness values (bottom). The maximum value is realized when $m = n = 9$. $b = 0.175$, $a = d = 0.04$, and $c = 0.57$, in which case $m_0 \approx 10$ and $m_1 \approx 15$. Since the actual number (m) of *HLA* loci is about 12, $m_0 < m < m_1$.

$$b \approx 1 - \exp(-2.3/m) \tag{3b}$$

(see below and Table 2). The average m per individuals in the *HLA* is about 12 (*A, B, C, E, F, G, DP, DNA/DOB, DM, DQ, DRA/DRB1+DRB3* or 5), so that we have $b \approx 0.17$ ($\theta_0 \approx 0.19$).

Secondly, thc fact that there indeed exist polymorphic *Mhc* loci imposes that

$$\frac{2\theta_1}{\theta_0} < a + c \text{ or } m_0 < m_1. \tag{4}$$

If $m > m_1$, none of m loci can be polymorphic. Therefore, the actual number (m) of *Mhc* loci must be smaller than m_1. Whether $m < m_0$ or $m > m_0$ is determined by g. If $g = 0.9$, then $m > m_0$. The fact that not all m loci are polymorphic requires also that

$$\frac{a}{1+a} + \frac{c}{1+c} < m\theta_1 \tag{5}$$

(see Table 1).

Thirdly, allo-polyploid *Xenopus Mhc*[27] provides useful information. Allo-tetraploid species, when first produced by hybridization, must have had twice as many functional *Mhc* loci as parental diploid species. The number of loci of diploid species is m and that of tetraploid species is $2m$. It is demonstrated that m such extra *Mhc* loci in allo-tetraploid species are rapidly suppressed or deleted[27]. This suggests strongly that $2m$ loci, even though all monomorphic, have substantial deleterious effects on fitness. Conversely speaking, deleting extra loci is actually advantageous. A conservative assumption is that any of $2m$ loci cannot be polymorphic, which leads to $f(n,2m) < f(0,2m)$ or $a + c < 2m\theta_1$ for $1 \leq n \leq 2m$. Therefore, we have

$$m\theta_1 < a + c < 2m\theta_1, \tag{6}$$

or equivalently $(a + c)/(2m) < \theta_1 < (a + c)/m$. In terms of $m_1 = (a + c)/\theta_1$, inequality (6) is rewritten as $m_1 < 2m$ and $m < m_1$. This sets a rather stringent

Table 2. Evaluation of parameter values.

m (ms)	θ_1	$b(\theta_0)$	$a + c \approx m(\theta_1 + s)$	m_0	m_1
5 (0.05)	0.04	0.37 (0.46)	0.25	4–5	6–7
8 (0.08)	0.04	0.25 (0.29)	0.40	6–7	10
12 (0.12)	0.04	0.17 (0.19)	0.60	10–11	15

Selection intensity for heterozygotes (s) and T-cell repertoire size ($1 - g$) are assumed to be 0.01 and 0.1, respectively.

relationship among the size of T-cell repertoire in heterozygotes determined by a, d, an excess (c) of the peptide repertoire, and the number of *Mhc* loci (m). Thus, even small changes in a, c, d and m may result in drastic changes in the evolutionary mode of *Mhc* genes.

Finally, there are attempts to estimate selective advantage of heterozygotes over homozygotes (s), based on the observed *Mhc* polymorphism in the human population.[5,28] The estimated s value is about 0.01 averaged over the *HLA-A*, *-B*, *-C*, *-DRB1* and *-DQB1* loci. The relationship between s and $f(n,m)$ is given by

$$1 - s = \frac{f(n,m)}{f(n+1,m)} = \frac{\{1 - a/(m+na)\}\{1 - c/(m+nc)\}}{1 - d}$$

at least for small n ($\leq m$). We note that $(a + c)/m - \theta_1$ must be in the same order as the small value of s: the smaller the number of functional (and heterozygous) loci, the larger the selection intensity (Table 2). We have

$$a + c \approx m(\theta_1 + \mathrm{s}). \qquad (7)$$

Substituting (7) for (6), we obtain $s < \theta_1$. This and $a + c > 2\theta_1/\theta_0$ lead to $m(1 + s/\theta_1) > m_0$. For a known m, θ_0 must be larger than $1/m$. Likewise, (5) and (7) yield $ms < a^2/(1 + a) + c^2/(1 + c)$.

The *Mhc* polymorphism should evolve if $a + c$ is much larger than d. A relatively small excess of blind spot stemming from different *Mhc* allelic products is the reason for the polymorphism. I would like to suggest that the a and d are in the same order of magnitude, since these are determined in the same thymus environment. The expression of host genes in infected cells is suppressed and foreign antigens are over-expressed.[1] Under this circumstance, Mhc can bind antigen peptides even with weak affinity. Such expansion of the antigen pool may increase the possibility that even slightly different Mhc molecules can present quite different antigens. Then it is the c that makes $a + c$ and d different by factor m.

The value of m_1 is rewritten as $m(1 + s/\theta_1)$. The m_1 is at most twice as large as m and becomes smaller as θ_1 increases from s. Assume that in the *HLA*, the optimum number of polymorphic loci is realized, which is about one half of the 12 loci. This requires that

$$s \approx \theta_1 \sqrt{\frac{ms}{2}}, \qquad (8)$$

which leads to $\theta_1 = 4s = 0.04$ and $a + c \approx 5ms = 0.6$ when $m = 12$ and $s = 0.01$. The expected value of m_1 is 15, suggesting that the *HLA* cannot evolve as a large multigene family with polymorphic loci. The actual number of nonoverlapping and functional *HLA* loci is close to m_1 so that even a small increase in the number of functional *HLA* loci in the genome is likely to prohibit the development of polymorphism. Also, it is clear that heterozygote advantages gained in positive selection and peptide presentation are diluted as the number of loci increases. After all, these advantages are at most fourfold ($a = c = 1$), whereas disadvantages of blind spot are unlimited. In actual immune systems, the net advantage for heterozygotes may be only marginal. In the evolutionary sense, however, 1% heterozygote advantage is sufficiently strong to maintain a large number of *Mhc* alleles for millions of years.[28,29]

A Speciation Mechanism

The reason why *Mhc* does not have a large number of functional copy genes despite the apparent advantage in presenting foreign peptides is that too many diverged *Mhc* genes result in too large a blind spot: The blind spot causes non-responsiveness and increases susceptibility to diseases. Suppose that some individuals have different alleles at all the functional class I and class II *Mhc* loci. These individuals may be produced by hybridization between antigenically diverged individuals. Figure 2 shows that the fitness function is maximized at a certain number of polymorphic loci and that hybrids heterozygous at more loci become less adapted. A necessary condition, however, is that an excess of blind spot (d) due to allelic products is substantial.

It is known that some allelic *Mhc* lineages have a long evolutionary history and are often shared by different species, the so-called trans-species polymorphism.[7] If such allelic lineages are not confined in some geographically subdivided populations, their role in speciation is unlikely. However, it is also observed that different species have nonoverlapping profiles of trans-specific allelic lineages.[30] Different *Mhc* alleles may happen to be distributed into geographically different populations and/or coevolve with local pathogens.

The number of *Mhc* loci depends on haplotypes, adding another source of *Mhc* diversity. Indiviuals having two different haplotypes may have too many *Mhc* loci. For instance, there are five *DR* haplotypes (*DRw8*, *DR1*, *DR2*, *DR3* and *DR4*) in the human population (Figure 3). The *DRB1* locus is shared by all haplotypes and has the same effect on the T-cell repertoire in all genotypes except its allelic differences. However, each of *DRB3*, *DRB4* and *DRB5* occurs only in one or two particular haplotypes, so that individuals such as *DR3*/*DR4* have two extra hemizygous loci compared with *DR1*/*DR1*. The reduction of fitness in *DR3*/*DR4* may be measured as $f(n,m+2)/f(n,m)$. This is proportional to $\exp(-2\theta_0)$, which is about 68% if $\theta_0 = 0.19$ (Figure 2). It is therefore possible that hybrids with different haplotypes are less defensible owing to the Mhc-restriction of T-cell clones. If we can evaluate the Mhc-restricted T-cell repertoire among various haplotypes, the possible role of the immune system played in speciation becomes clearer.

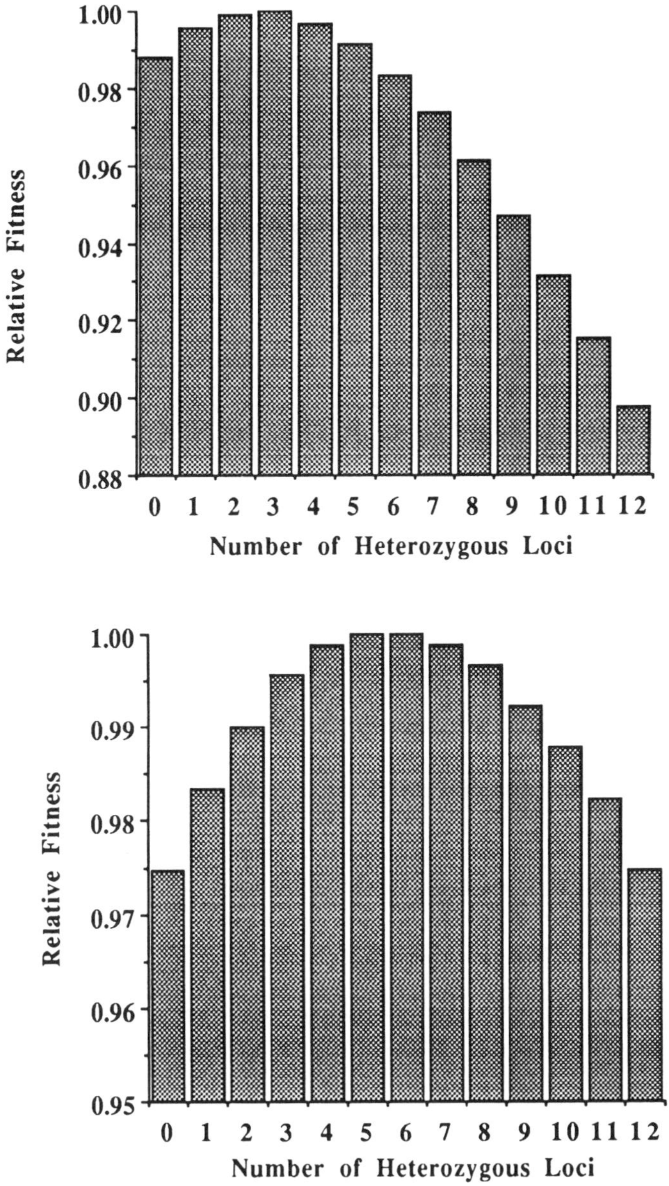

Figure 2. The fitness function against the number of polymorphic loci (n). $m = 12$ and $b = 0.175$. (a) $a = d = 0.04$, $c = 0.57$. (b) $a = d = 0.06$, $c = 0.803$.

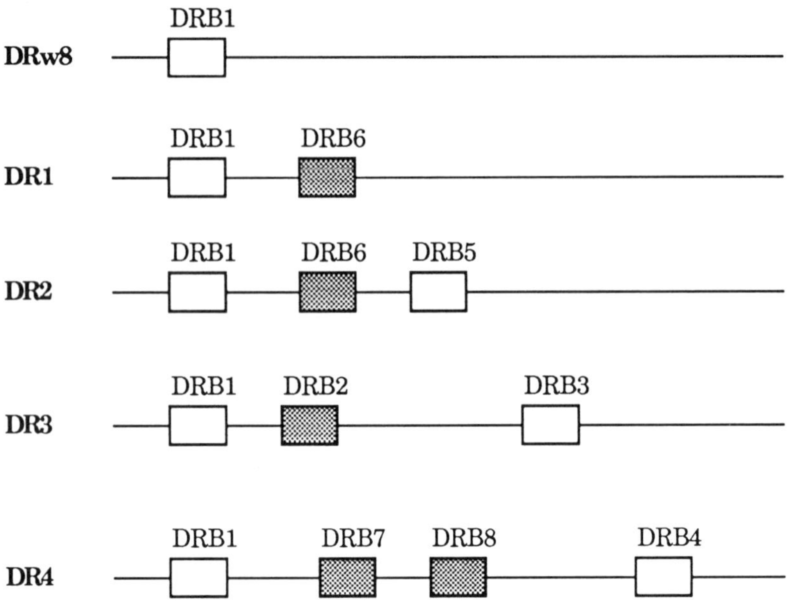

Figure 3. *DR* haplotyes in the *HLA*. *DRB2, 6, 7, 8* and *9* (not depicted) are pseudogenes.

It is particularly interesting to examine this possibility for species that went to rapid speciation processes. Cichlid fishes under explosive speciation in the East African Lakes Victoria, Malawi and Tanganika[31] appear to be one of the best organisms to be examined in relation to the *Mhc* gene organization and the extent of its polymorphism. The population that founded the species flock of Lake Malawi is suggested to have been highly polymorphic at its *Mhc* loci.[32] The *Mhc* loci are not only useful to infer the mode of speciation that happened some million years ago, but also interesting to examine their direct involvement in the speciation process.

In this paper, I have proposed a model of individual fitness associated with *Mhc* and used it to argue for the condition that balancing selection operates and a possible role of the immune system that plays in speciation. However, the dynamical aspect of the model is not exploired under the joint effect of mutation, unequal crossing over, genetic drift, and so on. This will be done elsewhere.

Acknowledgments

This paper is supported in part by the CIAR evolution program and in part by Grant 05044127 (Monbusho International Scientific Research Program: Joint

Research) and Grant 04304001 in the Ministry of Education, Science and Culture, Japan.

References

1. J. Klein. (1990). *Immunology*. (Blackwell Sci. Pub. Inc., Oxford).
2. H. M. Grey, A. Sette, & S. Buus. (1989). *Sci. Amer, Nov.*, 38–46.
3. A. L. Hughes, & M. Nei. (1988). *Nature*. **335**, 167–170.
4. S. A. Hill, C. E. M. Allsopp, D. Kwiatkowski, N. M. Anstey, P. Twumasi, P. A. Rowe, S. Bennet, D. Brewster, A. J. McMichael, & B. M. Greewood. (1991). *Nature*, **352**, 595–600.
5. N. Takahata, Y. Satta, & J. Klein. (1992). *Genetics*, **130**, 925–938.
6. L. E. Hood, I. L. Weissman, W. B. Wood, & J. H. Wilson. (1984). *Immunology*, (Benjamin/Cummings Pub. Co., Inc., Menlo Park).
7. J. Klein. (1986). *Natural History of the Major Histocompatibility Complex*. (John Wiley & Sons, New York).
8. R. D. Cambel, & J. Trowsdale. (1993). *Immunology Today*. **14**, 349–352.
9. V. Vincek, D. Nižetić, M. Golubić, M., F. Figueroa, E. Nevo, & J. Klein. (1987). *Mol. Evol. Biol.*, **4**, 483–491.
10. J. Klein, & C. Shönbach. (1991). (eds. B. G. Solheim, S. Ferrone, & E. Möller) 16–37 (Springer-Verlag, Heidelberg).
11. R. de Boer, & A. Perelson. (1993). *Proc. R. Soc. Lond.* **B**. **252**, 171–175.
12. S. J. O'Brien, D. E. Wildt, M. Bush, T. M. Caro, C. FitzGibbon, I. Aggundey, & R. E. Leakey. (1987). *Proc. Natl. Acad. Sci, USA*, **84**, 508–511.
13. J. B. S. Haldane. (1949). *La ricerca scientifica, supplemento* **19**, 2–11.
14. J. B. S. Haldane. (1942). *Annual of Eugenics* **11**, 333–340.
15. J. F. Crow. (1990). (ed. Dronamraju, K. R.) xi–xii (Garland, New York & London).
16. J. F. Crow. (1992). *Genetics*, **130**, 1–6.
17. M. Blackman, J. Kapper, & P. Marrack. (1990). *Science*, 248, 1335–1341.
18. H .V. Boehmer, & P. Kisielow. (1990). *Science*, **248**, 1369–1373.
19. Y. Takahama, E. W. Shores, & A. Singer. (1992). *Science*, **258**, 653–656.
20. P. J. Bjorkman, M. A. Saper, B. Samraoui, W. S. Bennett, J. L. Strominger, & D. C. Wiley. (1987). *Nature*, **329**, 512–518.
21. K. Falk, O. Rötzschke, S. Stevanović, G. Jung, & H.-G. Rammensee. (1991). *Nature*, **351**, 290–296).
22. H.-G. Rammensee, K. Falk, & O. Rötzschke. (1993). *Annu. Rev. Immunol.*, **11**, 213–244.
23. T. S. Jardetzky, W. S. Lane, R. A. Robinson, D. R. Madden, & D. C. Wiley. (1991). *Nature*, **353**, 326–329.

24. B. M. Carreno, R. W. Anderson, J. E. Coligan, & W. Biddison. (1990). *Proc. Natl. Acad. Sci, USA*, **87**, 3420–3424.

25. J. B. S. Haldane. (1955). *Proc. R. Soc. Lond.* **B**. **144**, 217–220.

26. H. von Boehmer. (1988). *Annu. Rev. Immunol.*, **6**, 309–326.

27. H. R. Kobel, & L. Du Pasquier. (1986). *Trends Genet.*, **2**,310–315.

28. J. Klein, Y. Satta, C. O'hUigin, & N. Takahata. (1993). *Annu. Rev. Immunol.*, **11**, 269–295.

29. N. Takahata. (1990). *Proc. Natl. Acad. Sci. USA*, **87**, 2419–2423.

30. J. Klein, & D. Klein (ed.). (1991). *Molecular Evolution of the Major Histocompatibility Complex*. (Springer-Verlag, Heidelberg).

31. C. Sturmbauer, & A. Meyer. (1992). *Nature* **358**, 578–579.

32. D. Klein, H. Ono, C. O'hUigin, V. Vincek, Goldschmidt & J. Klein. (1993). *Nature* **364**, 330–334.

33. J. Seger, J. (1988). *Proc. R. Soc. Lond.* **B**. **319**, 541–555.

Index